T0313717

"Embodies the tenet that one learns by practicing the craft in a supportive community where everyone continually strives for greater mastery. This book will be invaluable for undergraduate and graduate students in writing or journalism programs, regardless of their interest in the sciences. Students in the sciences who encounter this book may discover a second vocation. Highly recommended. All readers."

—*Choice Reviews*, American Library Association

"One of the best journalism books of all time."

—BookAuthority

"Buying this book, one is not merely purchasing a science writing guide. One is taking a seat at a community table. As the world faces threats from climate change and the coronavirus pandemic, science journalists, as first responders for truth, need every bit of that sustenance."

—*Journalism & Mass Communication Educator*

Praise for *The Craft of Science Writing*

"If I had to strip my go-to shelf of reporting and writing books down to a very few, this one would remain. *The Craft of Science Writing* offers valuable tips for any reporter or editor, of any experience level, covering any topic. In an era when facts are under assault, this book is especially welcome."

—Jacqui Banaszynski, Pulitzer Prize–winning reporter and editor, Knight Chair Professor Emerita at the Missouri School of Journalism

"Since 2010, *The Open Notebook* has served as an indispensable online guide to science writing. With *The Craft of Science Writing*, they've distilled those years of insights into an excellent survey of the field. You'll find the nuts and bolts here—how to read a scientific paper, how to craft a lede, and so on. But you'll also get a sense of what it's like to wake up every morning and face the delights and dread that this job brings."

—Carl Zimmer, journalist and author of *Life's Edge: The Search for What It Means to Be Alive*

"Listen up, journalism professors, students, and media professionals! You won't find a better guide to the art (and craft) of science writing than this book. *The Open Notebook* has produced a fascinating, informative, empathetic, and pragmatic tour through an increasingly vital set of skills. Whether you cover politics, education, national security, transportation, or business, the ability to write intelligently about science is now essential. Luckily for us, *TON* has distilled years of expert observation and insight to light the way."

—Jennifer Kahn, contributing writer for the *New York Times Magazine* and Narrative Program Lead, University of California, Berkeley Graduate School of Journalism

"Accessible, informative, and engaging, *The Craft of Science Writing* can serve as an excellent resource for both classroom instruction and self-study. A welcome contribution!"

—Barbara Gastel, MD, professor and Science & Technology Journalism Program coordinator, Texas A&M University

Praise for *The Open Notebook*

"I recommend *The Open Notebook* to every writer, not just science journalists. Their story dissections are amazing, their pitch database is a goldmine, and their profiles of other writers are the best way to score insider tips and/or feel less alone in the struggle."

—Nicola Twilley, author of *Frostbite: How Refrigeration Changed Our Food, Our Planet, and Ourselves*, cohost of the *Gastropod* podcast, and frequent contributor to the *New Yorker*

"Writing is a lonely art. The vast majority of the process, from coming up with ideas to structuring a piece, is carried out in solitude. When you start out, you barely have a clue about what you're doing, let alone what seasoned professionals get up to. Getting those pros to deconstruct their own methods for all to see is a massive boon to aspiring science writers."

—Ed Yong, Pulitzer Prize–winning science journalist and author of *An Immense World: How Animal Senses Reveal the Hidden Realms around Us*

"*The Open Notebook* is an incredibly generous project, a gift to the science writer community. The *TON* interviewers pick some of the most innovative and engaging recent science features and ask writers just the right questions. Everyone involved— the *TON* creators, the interviewers, the writers who reveal their methods—is helping teach the rest of us a master class in science writing."

—Laura Helmuth, editor-in-chief of *Scientific American* and past president of the National Association of Science Writers

"Most websites on media focus on ephemera: Who's in, who's out, mergers, layoffs, and corporate gossip. *The Open Notebook* is a rare, shining exception. Aiming to serve as a kind of *ars journalistica* for working science writers, the site lays bare the elements of craft that determine a story's lasting impact. By interviewing those who have distinguished themselves in the fields of science reporting, feature writing, blogging, and multimedia, *The Open Notebook* transcends hackneyed print-vs.-digital dichotomies to deliver tips, advice, and food for thought that can be directly put into practice in today's hyper-competitive freelance environment. The site also offers something that's harder to define but just as essential: a sense of collective ethics and joy in our hard work. I love *The Open Notebook*."

—Steve Silberman, journalist and author of *Neurotribes: The Legacy of Autism and the Future of Neurodiversity*

The Craft of Science Writing

The Craft of Science Writing

SELECTIONS FROM
THE OPEN NOTEBOOK
EXPANDED EDITION

Edited by Siri Carpenter

THE UNIVERSITY OF CHICAGO PRESS   Chicago and London

The University of Chicago Press, Chicago 60637
The University of Chicago Press, Ltd., London
© 2024 by The Open Notebook, Inc.
Published 2024
Printed and bound by CPI Group (UK) Ltd, Croydon, CR0 4YY

33 32 31 30 29 28 27 26 25 24   1 2 3 4 5

ISBN-13: 978-0-226-83027-8 (cloth)
ISBN-13: 978-0-226-83029-2 (paper)
ISBN-13: 978-0-226-83028-5 (ebook)
DOI: https://doi.org/10.7208/chicago/9780226830285.001.0001

Library of Congress Cataloging-in-Publication Data

Names: Carpenter, Siri, editor.
Title: The craft of science writing : selections from The Open
    Notebook / edited by Siri Carpenter.
Other titles: Open Notebook (Website)
Description: Expanded edition. | Chicago ; London : The University
    of Chicago Press, 2024. | Series: Chicago guides to writing,
    editing, and publishing | Includes index.
Identifiers: LCCN 2024013178 | ISBN 9780226830278 (cloth) |
    ISBN 9780226830292 (paperback) | ISBN 9780226830285 (ebook)
Subjects: LCSH: Science journalism. | Science news.
Classification: LCC PN4784.T3 C73 2024 | DDC 070.4/495—dc23/
    eng/20240329
LC record available at https://lccn.loc.gov/2024013178

Contents

Note to Readers

WE HAVE MADE only minor copyediting modifications to material that has been previously published at *The Open Notebook* or in the first edition of this book. Affiliations of the writers and editors whose voices are included in this volume were accurate when the articles were originally published, though many have since moved to other publications. In addition, we deeply regret the loss of our treasured colleagues David Corcoran and Sharon Begley, who contributed to *The Open Notebook* in numerous ways over the years, and whose wisdom you'll find in this book.

Introduction

Siri Carpenter

I SOMEHOW MANAGED to become a reporter and editor without ever taking a journalism class. And for a long time, I never quite felt like I knew what I was doing.

I was working on my PhD in social psychology when I stumbled on science writing as a career possibility. I was passionate about work in my field—the study of implicit bias—and I loved thinking and talking and reading about science and scholarship. But the prospect of drilling ever deeper into a single academic field also felt confining. This discomfort with my chosen future nagged at me for months, but I didn't quite know what to do about it.

Then, while reading the *New York Times* science section one Tuesday over lunch in my department's graduate student lounge, I realized that the people who wrote these articles, whose minds were given license to flit from one fascinating topic to another—that this was their job. That *science writing* was a career. That it could be my career. And that I wanted that career. I wanted it badly.

I was fortunate to have advisors, internship supervisors, and mentors who set me on a path toward that goal. But even after a number of years on that course, I sometimes questioned whether I was cut out to be a writer. I found reporting exhilarating—and best of all, I was being paid to learn. But then there was writing. Writing had its moments of exhilaration, too, but it could also be excruciating.

So when I read a riveting science story, I was filled with admiration and inspiration—but also envy and a sense of inferiority. What did these writers know that I didn't? How, say, did *Science*'s David Grimm get to the bottom of a story about scientists using fallout from atomic-bomb tests to solve crimes? How did award-winning *Science News* reporter Tina Saey decide what to include in her wide-ranging story about the function of sleep? How did Roberta Kwok, reporting for *Nature*, figure out how to build a compelling narrative about an asteroid falling to Earth? I longed to ask them a thousand questions, to get down to the nitty-gritty details of how they did it, to peek inside their reporters' notebooks.

When I discovered that my friend and fellow science writer Jeanne Erdmann harbored the same kinds of doubts that I did, we decided to do something about it. Maybe, we thought, these science writers and others would be willing to tell us how they wrote the stories we so admired if we wrote up the interviews and published them on a website.

Thus, in 2010, *The Open Notebook* was born.

Our ambitions were modest, at first. Jeanne and I imagined that, over time, we'd talk with a dozen or so writers and that, collectively, those interviews would point the way toward the Lost City of Amazing Science Journalism. But before long we recognized the potency of these discussions. We realized that just as there is no one right path into science journalism, there is no single right way to find, pitch, report, or write a science story. And when we shared the initial interviews with others, we were met with an overwhelming response. It was clear our fellow science writers also hungered to glimpse others' methods and to hone their own skills.

Since those early days, *TON* has expanded and fine-tuned its mission. Above all, we believe that outstanding craftsmanship still matters. We aim to guide science writers in all aspects of the craft: finding and honing story ideas, pitching to editors, creating a reporting plan, identifying and interviewing sources, organizing notes, and then—finally—coping with that inevitable moment when the office has been thoroughly cleaned, the sock drawer Marie Kondoed, and there's nothing left to do but start writing. We also offer guidance on the next steps: structuring stories, crafting compelling and memorable narratives, rewriting (and rewriting and rewriting), and collaborating with editors to sculpt stories into their final form.

The Open Notebook also looks beyond the craft itself. Writing can be lonely, and the voices inside your head can be relentless with their questions: "Why is this so hard for you?" "If you're so good, why aren't you earning more money?" "Have you considered that maybe you're just not cut out for this?" "How can everyone else pump out vibrant, clever, erudite, impactful, funny, moving stories when you can't even reliably remember how to spell *impostor*?" Part of what drives these feelings of inadequacy is the inconvenient fact that we can't see inside other people's minds. When we read others' work, work we admire, we see only the gleaming end product—we don't see the false starts, the hours of staring at (and avoiding) a blank page, the crumpled paper in the wastebasket. So, woven into everything we do at *The Open Notebook* is the goal of offering a glimpse behind the curtain: to reveal not only the moments of inspiration and joy that science writing can bring, but also the toil, the self-doubt, the dead ends. The *true* stories behind the stories.

We also began to create other tools and resources for science writers. With a grant from the National Association of Science Writers, we commissioned articles on aspects of the craft of science writing. To help our readers better understand the dark magic of writing pitch letters, we created a Pitch Database, which now contains more than 300 successful news and feature pitch letters. We started three regular sections: "Ask *TON*," an advice column; "Office Hours," which offers insights from journalism instructors; and "A Day in the Life," mini-profiles focused on the working lives of respected science writers. And eventually, with a grant from the Burroughs Wellcome Fund, we started an early-career fellowship program, in which fellows report and write stories for *The Open Notebook* with the guidance of a mentor. The program has evolved into a community of young science writers and the mentors who worked with them.

By now, *TON* has published more than 600 articles online: dozens of writer interviews, annotations of award-winning science stories, writer profiles, and deeply reported articles on many, many aspects of the craft of science journalism—how to find sources for science news stories, how to report on disability, how to make FOIA requests, how to interview sources who've experienced trauma, how to annotate articles for fact-checking . . . and a whole lot more.

Over the years, people asked us whether *The Open Notebook* might publish a compilation of *TON* articles—a book they could hold in their hands and dog-ear and scribble in; a guided tour through some of the core themes we've proved; a record of a collective dialogue that would give the voices in their head a stern talking-to.

The first edition of *The Craft of Science Writing*, published in February 2020, was that guide, compiling articles published over the course of several years. I could not have been more thrilled at how the science-writing community embraced the book. Aspiring science writers have used it to find a toehold in the industry. Journalism teachers have built courses around it. Experienced science journalists and science communicators at universities and other institutions have used it to sharpen their skills. Local and general-assignment journalists have relied on it to better understand how to bring science into their beats. Scientists have turned to it to hone their ability to communicate with the general public. And others who care about the role of science in society have used it to become more critical consumers of scientific information.

That first edition came out just before COVID-19 was declared a pandemic. At its release, we couldn't know how profoundly the profession would be tested over the coming years, nor how much the public would soon depend on science journalism to understand fast-changing scientific developments and make basic, day-to-day decisions about whether and how to go to school and work, to travel, to gather with loved ones.

The pandemic magnified public awareness of the countless ways science permeates our lives and the degree to which we rely on scientific information to navigate daily life. At the same time, it revealed a deep public distrust of science and scientists—a distrust often fueled by those who would profit from spreading misinformation and disinformation. Complicating matters further was the lack of clear, consistent, honest guidance from public institutions, including government health agencies. The uncertainty and confusion that pervaded responses to the pandemic and reactions to those responses thus underscored that science is a human endeavor, and a fallible one, not free from political influence.

So for this second, expanded edition of *The Craft of Science Writing*, we've added nine new essays. Some focus on equipping science writers with essential skills, including how to establish a beat in science journalism, how to find and use quotes, how to critically evaluate scientific claims, how to use social media for reporting on health, and—a skill whose importance was highlighted during the early days of the pandemic—how to cover unvetted preprint manuscripts. Reflecting the continuing evolution of science writing beyond an endeavor involving words on a page, we've added an essay on doing data journalism. Finally, we've sampled from our extensive Diverse Voices in Science Journalism collection, including essays on eradicating ableist language from science stories, working with a sensitivity reader, and breaking into English-language media for speakers of other languages.

The new selections were informed, in part, by feedback from readers of the first edition—including early-career journalists, established professionals, science-writing instructors, and their students. As we explored the idea of a second edition, we heard from those who had read the first edition that they were eager for more: about the fundamentals of reporting and writing well; about covering science with equity and inclusion in mind; about using emerging tools and challenges in the field. There's so much more we could have included. In just the four years since the first edition came out, we've published some 200 articles at *The Open Notebook*, on topics ranging from doing multimedia and photojournalism projects, to covering AI, to writing opinion pieces as a journalist, to writing science explainers for local audiences, and so much more. We can't fit it all in just one volume, but we've tried to give you a curated sampling that we hope will leave you hungry for more.

The volume is organized into five parts, structured to take readers on a journey through the profession, from exploring what it means to be a science journalist and taking first steps toward becoming one; to finding story ideas and convincing editors to assign them; to reporting out all the important details; to making your stories sing; and finally, to deepening your understanding and expertise. I've designed the book so that you can either read it all in order, from start to finish, or flip to whichever essays call to you. It's not a murder mystery; reading the ending first won't spoil anything. Choose your own adventure!

By the time you finish this book—the how-tos, the wherefores, the in-depth conversations about covering infectious disease and documenting panda sex—I hope that you'll have started to view reporting and writing (and revising, and revising again) as a set of skills that are challenging, variable, even maddening—but absolutely learnable.

I want you to walk away from this book with new insights about how to find a story idea, how to sell that idea to an editor, how to rev up and start reporting, and how to get the most out of that reporting to build the best possible story. I want you to be able to pick up a scientific paper and read it with a journalistic eye, to delve into the statistics and think critically about the conclusions and claims that are made from them. I want to buoy you, to help you feel confident in muddling through the writing part: how to start a story, how to tell it not-so-elegantly in the first draft, how to delve back in to make it elegant, and how to finish it.

Every writer before you has struggled with many or all of the same challenges. This book is here to show you that you are not alone, that every barrier you encounter is one that others have encountered before you. In this volume, some of the writers I admire most describe their challenges—and their solutions—so that their experiences might help you move forward, too.

More than anything, I hope their words will help you understand—will help you know in your bones—that you are not an impostor. And for the record: It can be spelled either "impostor" or "imposter." Either is fine. Really.

PART 1

Who Is a Science Journalist and How Do You Become One?

was refining her craft, as she did when reporting on Ebola outbreaks in Sierra Leone. She spoke to a survivor of Ebola months after their initial interview, and he "told me that after he relived the experience while talking with me, he felt terrible for days and didn't leave the house." She realized she'd made a "terrible mistake" in her interviews, focusing only on getting the information she needed and not on the impact she had on one another in

AS AN OUTSIDER to science journalism, it's easy to imagine that journalists writing about science must have some background that you don't have. They must have training or experience as scientists or journalists or both, or specialize in science journalism, or live in major media centers like New York or Washington, or write in perfect, sparkling English. Somehow, they must possess *some* special sauce—right?

Wrong. The good news is that there are lots of paths into science journalism, and each comes with its own advantages and its own pitfalls. The trick is to learn to exploit your inherent strengths and shore up—and work around—your weaknesses. That's what this section is all about.

Yes, some science journalists have PhDs and years of research experience. Their depth of knowledge about their field may allow them to tell stories whose importance might elude other journalists—but at the same time, they enter the field without knowing the conventions of journalism, and their very knowledge may make it a challenge to put themselves in the mind of a general reader for whom the whole subject is new. And yes, some science journalists have training or years of experience in journalism—but they must learn to distinguish science that is solid from science that is iffy, to find the right experts for a story, to scrutinize statistics, and more. General assignment reporters who don't specialize in science still have to make sense of science for their readers when they cover a public health emergency, or local environmental issues, or transportation, or public housing, or immigration. And while a resident of Bangkok or Medellín or Kampala might understand scientific challenges and breakthroughs in their country with unrivaled depth and nuance, they may need some tips and tricks to break into English-language publications.

But the most important tools are curiosity, empathy, a commitment to getting the story right, and a dedication to continuing to hone your skills. Even an extremely experienced reporter like Amy Maxmen is always refining her craft, as she did when reporting on Ebola outbreaks in Sierra Leone. She spoke to a survivor of Ebola months after their initial interview, and he "told me that after he relived the experience while talking with me, he felt terrible for days and didn't leave the house." She realized she'd made a "terrible mistake" in her interviews, focusing only on getting the information she needed and not on the impact she had on people who had been through horrific traumas. "I had no idea that I was causing this pain, and the only way to fix the situation now is to not repeat that mistake, and to share the story so that other journalists don't do the same," Maxmen says.

And that realization lies at the core of this book: We can support one another in improving our craft and better serving our readers, our sources, and the communities they're part of.

Siri Carpenter

1 How to Use Reporting Skills from Any Beat for Science Journalism

Aneri Pattani

AS A BUSINESS and lifestyle reporter, Arizona-based freelance journalist Alaina Levine never expected a newspaper ad about spiders to play a key role in her journalism career.

But in 2012, it landed her a story with *Scientific American*.

Levine had been a columnist and freelancer for *Inside Tucson Business* for years, writing about business and public relations, as well as food and nightlife. As part of that job, she made a point of reading local newspapers to stay informed about events in the area, brainstorm ideas, and think of potential new sources. In July 2012, she spotted "a speck of an announcement" about the American Tarantula Society conference, which was to be held in Tucson that year. Although the topic had little connection to her usual beats, it sparked her curiosity and she decided to attend.

At the conference, she heard a graduate student giving a talk about her research in forensic entomology, the burgeoning field studying the succession of insects that feast on corpses at crime scenes. It was something Levine had never heard of before, but she found the presentation fascinating and asked the student to meet her for coffee. That simple coffee date resulted in a successful pitch to *Scientific American* about how the student's research was upending the field of crime scene investigation. Levine says that science story—and other opportunities in science journalism that followed—stemmed directly from her focus on keeping up with the local business beat.

It's a pattern Levine says has repeated many times throughout her career. Writing about business taught her how to break through the army of public relations specialists surrounding prominent executives like Elon Musk to get access for in-depth profiles—a skill she now uses in covering technology. And writing columns that were meant to give people advice in a brief space helped her focus on a few key takeaways—a skill she now puts to use distilling complex research or the most significant impacts of a new scientific discovery.

"Because of that exposure, I feel I'm a stronger science writer," Levine says.

And she's not alone. Many science reporters get their start in other beats, including crime, business, fashion, or politics. And many of them agree that the skills they developed in those beats have made them better equipped—and sometimes uniquely positioned—to report on science.

That's what Abigail Foerstner believes. She's codirector of the health, environment, and science graduate specialization at Northwestern University's Medill School of Journalism. She's been teaching science journalism for more than a decade, often to graduate students who have experience in many different beats.

Science stories touch every aspect of life, Foerstner says. Understanding the political arena can help reporters cover climate change policy. Understanding the business of research funding or the roadblocks to product development can help reporters cover new scientific advances with nuance and healthy skepticism.

"Anything you do could help inform science writing," Foerstner says.

Meredith Rutland Bauer

PREVIOUS BEAT: General assignment
CURRENT POSITION: Freelance science and technology writer
When covering the devastation of Hurricane Harvey in Texas in August 2017, Meredith Rutland Bauer found the son of a woman with type 2 diabetes whose house had flooded. The woman couldn't get out and the son was desperate to find her help. Bauer knew her story depended on extracting details from the son about the condition of the house and how much medication his mother had remaining. But she also knew that information could only come from a probing and personal conversation, which would be difficult in a time of crisis. The way she convinced him to open up to her, she says, came down to skills she'd picked up as a general assignment reporter.

Before going into science writing, Bauer worked at the *Florida Times-*

Union in Jacksonville, covering everything from school board meetings to the shooting of an unarmed Black teen. She also worked on the traditional training ground for young reporters: obituaries and memorials.

Working on obituaries in particular "gave me a really good sense of how to approach topics that are sensitive and how to approach people when they're under stress," Bauer says. "That's something that's applicable to covering a crime at 3:00 a.m. or a story on ecosystem damage."

She learned to be honest with sources and not give false hope about what her article could do for them, but still explain why she felt their story was important to tell. She applied that same tactic to speak with the diabetic woman's son during the flooding in Houston. She didn't promise her article could help his mother get more insulin, but she did explain that the story would help people outside of Texas understand how dire the circumstances were for people stranded during Harvey.

As a science journalist, that tact has become invaluable for her. "Being able to cover natural disasters in an empathetic way to get this story out there is really important," she says, "especially as climate change creates a whole host of disasters."

What advice do you have for someone in general assignment who wants to do science journalism?

MEREDITH RUTLAND BAUER: Focus on strong, powerful stories in general, and show that you can over-deliver on your articles. If you have to take an extra assignment that you do on a weekend or stay at work late to find a way to plug a scientific study into your beat, then do it. It's okay if it's difficult at first. But know that it's very possible to make the transition to science.

Christina Couch

PREVIOUS BEAT: Business/finance
CURRENT POSITION: Freelance science writer
One reason Christina Couch loves covering technology is because it's an optimistic field. Unlike a lot of news that often focuses on what's going wrong, tech reporters get to chronicle advances that can improve

people's lives. But sometimes that type of reporting goes too far into hype, she says.

"A lot of tech reporting is very breathless," Couch says, describing the prevalence of articles proclaiming tech fixes to global poverty or world hunger. "But understanding the very concrete reasons why [much-hyped technologies] often never pan out is important to me."

According to Couch, her focus on those concrete reasons stems from her time as a business reporter. She spent more than seven years writing about student loans, personal finance, and small business before switching to science.

That experience, especially covering venture capital funding, taught her to develop a healthy level of skepticism, she says. Now, as a science reporter, she doesn't just write about every new tech startup's plan to "disrupt" the way we live. Instead, she researches the company's venture capital funding, the founder's past projects, and where the product falls in the development pipeline. She knows to ask about a company's runway (how long it can survive if income and expenses stay constant) and burn rate (the rate at which a new company is spending its venture capital before generating positive cash flow) to determine how much confidence investors have in the startup.

"I'm not 100 percent sure I would be focused on those types of issues if I didn't come from a business background," she says.

What advice do you have for someone in business reporting who wants to do science journalism?

CHRISTINA COUCH: I would get very familiar with how products are developed, whether they are tech-related or not. Really familiarize yourself with venture capital funding and why businesses fail. Also, learning to write a basic human profile is tremendously useful. A lot of times when we talk about science writing, we just talk about the science portion, but science writing comes with other parts like human interest and features too. Profiling a company is a great way to build those skills.

Donald G. McNeil Jr.

PREVIOUS BEAT: Foreign correspondent
CURRENT POSITION: *New York Times* global health reporter

No matter where in the world he's reporting, there's one question Donald G. McNeil Jr. always likes to ask his interviewees: What do you think made you sick?

It seems straightforward, but McNeil, a global health reporter for the *New York Times*, says he's gotten some surprising answers over the years. People sometimes tell him their neighbor cursed them, or that their ancestors are unhappy.

It's a question McNeil says he learned to ask from his past life as a foreign correspondent. Traveling through more than 50 countries reporting on topics as varied as the pope's visit to South Africa or a coup in Burundi, McNeil learned about different cultures, thought patterns, and belief systems. It taught him that not everyone processes life events in the same way Americans do, whether those events are wars, elections, or illnesses.

"You learn to ask questions you wouldn't ask otherwise," he says. If you're visiting a new country for the first time, "it might not occur to you to ask someone, 'How do you think you got this illness?'"

Science writers often assume people act rationally, McNeil says. They spend significant amounts of time interacting with researchers and academics, so it can be easy to default to a scientific explanation for everything. But that's not always the full story.

"You can't act as if everybody is a patient in America," he says. "You can't assume. You have to ask a lot of questions and get that into the story."

On a trip to South Africa to report on AIDS, for example, McNeil asked doctors whether their patients preferred traditional remedies. It turned out one of the doctors actually kept in touch with a network of traditional healers. He knew his patients—even the college-educated ones—would often follow up a visit with him by seeing a healer. The doctor tried to coordinate with the healers' remedies, so any herbs or other substances they suggested wouldn't react negatively with medications he was prescribing.

McNeil says he never would have gotten that story had he not consciously thought to ask that extra question. That's his takeaway from years in the field: "Ask enough questions so that you really understand *people*, not just the science."

What advice do you have for someone who's a foreign correspondent and wants to do science journalism?

DONALD G. MCNEIL JR.: No matter where you are, a lot of reporting is just remembering never to leave out the human element. Learn how to get people talking by asking what they had for breakfast or what type of work their parents did. Those are useful skills that'll help you really understand them.

Rebecca Boyle

PREVIOUS BEAT: Crime
CURRENT POSITION: Freelance science writer
One of the most memorable cases Rebecca Boyle covered as a crime reporter at the *Greeley Tribune* in Colorado was a double homicide where the prosecution was seeking capital punishment.

At first, however, there was a mistrial, which led to an extensive process to start a second trial. It involved a number of motions, pretrial hearings, and evidentiary proceedings. While some would consider that complicated legal process dull, it fascinated Boyle.

"To me, it was really important to go through that," she says. "What did it mean for someone to go through a murder trial? I wanted to understand the process."

That perspective has served her well as a science reporter. She often sees research papers, which present the final results of a study, but Boyle likes to ask how the researchers arrived at those results. And she makes sure to include those details in her articles.

"I think many journalists don't write about it, but you need that background to grasp what that research means to someone's career and why it's a big deal," Boyle says.

Writing about crime also helped Boyle learn about scientific procedures. She used the forensic evidence brought into court as a vehicle to learn about techniques like DNA testing, fingerprint analysis, and ballistics.

"I remember being like, 'Can you really tell which bullet came from which gun based on these stripes?'" she recalls. "So I started learning more about that."

It taught her not only about the science behind the crimes, but also how to write about methodologies in a way that's accessible for most readers. That often involves using examples and recognizing which details are extraneous, she says.

What advice do you have for someone in crime reporting who wants to do science journalism?

REBECCA BOYLE: Incorporate science writing into your everyday beat writing. Science is not separate from other topics. It's basically the process of figuring out how and why things work. Really, so is being a journalist. You can figure out how a person gets incarcerated for his entire life, or figure out the process behind a new paper on epigenetics. You can write about science in the context of anything. You just have to think about it creatively.

Ashley Berg

PREVIOUS BEAT: Fashion/beauty
CURRENT POSITION: Medical writer for Rosemont Media

Most people probably don't think fashion relates to science. But Ashley Berg found a story to prove them wrong.

Berg is a science writer for Rosemont Media, where she creates web content about elective medical procedures. But she comes from a background in fashion and beauty journalism. It's that training that helped her spot the unique story she's currently working on: about the pattern makers who work at NASA.

In fashion, a pattern maker draws a map of how to piece together a garment. At NASA, these employees create the fabric shields that cover satellites and protect them from burning up upon returning to Earth's atmosphere.

Berg says when she discovered that NASA employed pattern makers, "It was like a lightbulb went off." She realized few other people would think to write a story around this, but given her background in fashion, she could flesh out a story and ask questions that the average science writer might not think of, like discussing seam allowances (extra material that is left where the edges will go together and be sealed) and lining (material that is often used as an additional layer and, in the case of satellites, could be protective).

Berg's fashion reporting background has also shaped her writing style.

"Fashion is very much about glitz, glamour, and narrative," she says. "No one wants to read a blog about fabric types. They want stories built around the fashion."

The same is true for medical writing, Berg says. While some of her colleagues write in a more straightforward manner, emphasizing the dry facts, Berg prefers a more literary approach. "I like to create a little bit more of a story," she says. "I want to have more depth than simply providing the facts." It pushes her to find human narratives to illustrate the medical procedures she's explaining.

What advice do you have for someone in fashion reporting who wants to do science journalism?

ASHLEY BERG: Research is the most important skill to develop. Read other science writers. Learn from their work. Learn how they turn a nice phrase or identify an intriguing topic. If you can do that, then you can swiftly realize how to apply that to a new beat. Also, pick the brains of someone in medical or science writing. Try a science story in your free time and ask for their opinion on it.

Seth Mnookin

PREVIOUS BEAT: Politics and media
CURRENT POSITION: Freelance journalist, author, and codirector of the graduate program in science writing at MIT

One thing that politics and science have in common, says Seth Mnookin, is that no one can agree. That means reporters are often left trying to figure out what the truth is or which view has the most evidence on its side.

Today, Mnookin is a science reporter and book author, but during the 2000 presidential campaign he was covering media and, in particular, the arena of political reporting for the now-defunct magazine *Brill's Content*. The experience taught him how to decipher what information was worth reporting when there's disagreement.

"A crucial thing then was finding people who I could trust," he says. "People who I could call even if they weren't directly related to the story and say, 'I'm hearing this. Does that sound crazy to you?'"

It's a tactic he uses all the time in science writing now, calling researchers in a field to ask about a study or gauge whether a new treatment lives up to the hype.

He's also learned never to assume anything. On the campaign trail, Mnookin was a rookie political reporter surrounded by journalists who'd been covering politics for years.

"They'd talk about campaign managers from five campaigns ago," he recalls. "I felt like I had no idea what was going on."

At first, he was self-conscious about his inexperience, but he eventually realized he could use it to his advantage. "I didn't know enough to know the questions I was supposed to feel too dumb to ask," he says. "And in asking those questions, I often got the most revealing answers."

It's something that's become a mantra for him ever since. At MIT, he teaches his students not to be intimidated by scientists, but to ask the questions their readers may have. Sometimes Mnookin will even ask a question he knows the answer to, simply because the way scientists respond can reveal something about how they think or shed new light on the topic.

"I've found it's really easy to feel dumb when talking to a scientist," he says. "But for me, it's been really important to embrace that."

What advice do you have for someone reporting on politics or media who wants to do science journalism?

SETH MNOOKIN: The most important thing you can do is look for a good editor. I'm someone who believes a good editor can cure all ills. So if you don't have a science background, but you know of a science editor that really knows their stuff, try to work with them. That way, when you're getting your feet wet, you have someone there to make sure you don't fall on your face. Because when you start off, it can be hard to know what those potential screwup situations are.

2 Trading the Pipette for the Pen: Transitioning from Science to Science Writing

Julia Rosen

SOMETIME DURING MY PhD studies, I had an epiphany: I liked learning about science more than I liked doing it. Although I had excelled in science classes as an undergraduate, I was unprepared for the drudgery of lab work, and the funnel of ever narrower research questions that felt ever more removed from the questions that motivated me at the outset.

By the end of my third year, all I wanted to do—while mindlessly running samples, while putting off writing computer code—was read popular science. Stories about diverse subjects, from neuroscience to astronomy, captivated me as much as my own work on ice cores. I had always loved to write, and I wondered if, perhaps, I should try my hand at science writing.

This isn't an unfamiliar tale, according to Rob Irion, who directs the graduate Science Communication Program at the University of California, Santa Cruz. Many of his applicants, especially those who have completed or are working on PhDs, say that instead of reading the literature in their own field, they "spend a lot of time in their department seminar rooms devouring everything else, and finding that they are really interested in all of it," Irion says. This might be the closest thing to a litmus test to determine if you, like I and many others, might be a science writer trapped in a scientist's career path.

Making the transition to science writing can seem intimidating; it certainly requires determination and effort. But taking a few critical steps, like networking and honing your writing skills, can help make a scientist's next experiment on the page a successful one.

Get the Inside Scoop

If there's one thing science teaches you, it's how to learn something new. This time, instead of exploring how neurons fire or why volcanoes explode, you need to learn what science writers actually do and how they do it.

A good place to start is a literature review, including Carl Zimmer's note to beginning writers and Ed Yong's collection of science-writer origin stories. You could also study Irion's recent journal article, geared specifically toward scientists with an interest in communication. And you should pick up a copy of the excellent *Science Writers' Handbook.*

Next, do some fieldwork and meet some science writers in their natural environment. Irion recommends joining a local science-writing group (most major metropolitan areas have one) or reaching out to the people who handle media relations (often called communications, outreach, or public information officers) at your institution or company. They can give you a sense of the differences between public relations and journalism, and they often have experience with both.

Irion also recommends joining the National Association of Science Writers (NASW; new student memberships are only $20 for the first year) and attending the annual ScienceWriters meeting, a collaboration between NASW and the Council for the Advancement of Science Writing. There, you can talk with dozens of writers. "You will walk away with a really different perspective about what people do in this profession," Irion says. "It's amazingly catalytic."

You can also reach out to individual writers who live in your area (you can find them on the NASW registry) or who cover subjects that interest you. "I can't tell you how many people I have emailed out of the blue," says Bethany Brookshire, a pharmacologist-turned-writer now at *Science News.* "They have been so kind in response."

If you can, grab a coffee or schedule an informal lunch date with these writers and ask about their career paths and what they do on a daily basis. These conversations helped demystify science writing for me, and I always came away laden with good advice and plenty of leads on possible internships, fellowships, and places to pitch.

That's also what happened to freelancer Helen Shen—a former neuroscientist based in the San Francisco Bay area. She sat down with Emily Singer of *Quanta Magazine,* who happened to be the sister of a friend. Singer told Shen about internship and graduate-study opportunities that she eventually pursued.

Shen's experience is a good reminder not to write off your personal network when you're launching a new career. "You might be surprised to find out how many people know someone in science journalism," she says.

Put Pen to Paper

Of course, talking about writing only gets you so far. At some point, you have to put pen to paper and see if you actually like doing it. Writing classes can be a great way to test the waters, says Jill Adams, a pharmacologist who worked in industry and then as a non-tenure-track faculty member at New York University for eight years before jumping into science writing. (Adams took an online nonfiction course offered by the Gotham Writers Workshop; many other organizations and community colleges offer similar opportunities.)

If you are still in grad school, look no further than your university's English department or student newspaper. "You are kind of in this cozy, safe environment to learn and get that experience and make those connections without floating in the real world," says Eric Hand, a reporter for *Science* who took classes in journalism and covered the arts for the *Stanford Daily* while pursuing a PhD in geophysics. Students can also explore writing opportunities with their school's press office and alumni magazine.

Science-writing workshops are another way to explore the profession and start making connections. Shreya Dasgupta, a wildlife researcher based in Bangalore, India, attended a two-week workshop, where she gathered tips and met mentors and peers that helped her launch her freelance career. I had a similar experience at the Santa Fe Science Writing Workshop, a week-long event with outstanding instructors who teach the basics of science writing and provide feedback on your work.

Alternatively, some prospective writers make their own opportunities online. In 2008, Brookshire started blogging under the pseudonym Scicurious after meeting a writer at *Scientific American* who encouraged her to develop her skills that way. "From that point on, I wrote two to three pieces per week, all through my PhD and again through my postdoc," she says.

She built a following and discovered that she loved writing—enough to turn down an academic position after her postdoc. "By then I had already been writing online for about five years, and I'd just become obsessed with the joy of sharing science with the world."

Hone Your Chops

If talking to professional writers and dabbling in writing only fans the flames, it's time to sharpen your writing and reporting skills. Every scientist goes about this in a different way, but in general there are three strategies, pursued alone or in combination: internships, science-writing graduate programs, and the DIY approach.

Many magazines and newspapers offer internships, and NASW hosts an internship fair at the annual meeting of the American Association for the Advancement of Science (AAAS). One opportunity that's particularly relevant for those with science backgrounds is the AAAS Mass Media Fellowship, which places science students at media outlets like NPR, *National Geographic*, and, in my case, the *Los Angeles Times*, for 10 busy weeks.

Internships like this provide a crash course in journalism for scientists. And for those who want to continue to write after the internship is over, the stories they've written can be used as clips to land their next journalism job—or to pursue more formal journalism training. After Shen finished her AAAS fellowship at the *Philadelphia Inquirer*, she enrolled in the year-long Science Communication Program at the University of California, Santa Cruz.

Graduate programs in science, health, or environmental journalism can also be found at many other universities, including MIT, the University of Colorado at Boulder, Boston University, Johns Hopkins University, New York University, and the University of Georgia, and many involve a mix of coursework and internships.

Although Irion stresses that people should take the path that's right for them—and carefully weigh the cost of tuition—he thinks that "an organized training program provides you with a very deep tool kit" that accelerates students' learning. "I see students from my program and elsewhere getting good jobs after one internship," Irion says.

However, many science writers have built successful careers without participating in these programs. "I have never felt at a disadvantage," says freelancer Jennifer Carpenter, who was doing her PhD in evolutionary genetics at the University of Edinburgh when she started writing press releases for the European Science Foundation. She went on to do internships at the BBC and *Science*.

Along the way, she took charge of her own education, reaching out to potential mentors and studying the craft of journalism. "I read the *Columbia Journalism Review* pretty religiously," she says, and listened to NPR's *On the Media* and BBC Radio's *Feedback*, which featured commentary and criticism of the BBC's programming. Carpenter also read—and continues to reread—codes of ethics to keep up on the rules of the profession, and tried to absorb wisdom from her editors.

Face the Challenges

Scientists who sidestep the ivory tower for science writing—or any other career—often face barriers to leaving. Grad students are often conditioned to think anything other than the tenure track is a failure—not to mention a disappointment to their advisors, their friends and family, or themselves. "Many people, including me, go into academia having never wanted anything else in their lives," Brookshire says.

But there are more challenges scientists should be prepared to face as they continue their metamorphosis. For example, many struggle to connect with lay audiences and have to recalibrate their sense of what's newsworthy. Adams says she used to use her husband—a nonscientist—as a sounding board for story ideas. Doing so helped her realize which subjects would interest only a specialized audience and reminded her that certain concepts she took for granted were foreign to many readers.

Many scientists also suffer from a form of perfectionism—and an almost pathological attention to detail—that can be debilitating in writing. "I observe a lot of mental struggles to realize that not every article that you produce is going to be the be-all and end-all of articles on subject X," Irion says. You can do a good job without agonizing, he says. "In fact, you have to."

That's especially true when adjusting to journalism's short deadlines after working at the glacial pace of academic research for so long. Shen was shocked by the fast turnaround times at the *Philadelphia Inquirer*. "At the newspaper, as soon as your editor tells you that you should work on something, you're already late."

Many scientists also need to shed their timid habits—by, for example, learning to conduct interviews over the phone rather than email, and asking tough questions that sources may not want to answer. They also may carry with them an engrained sense of caution when it comes to the wider implications of research; as journalists, however, it's their job to show readers why science matters on a larger scale.

But while having a scientific background carries particular challenges, it also confers particular benefits. Early on, Adams already knew to ask scientists about the process that led to their new results and the hitches that almost certainly occurred. "I wasn't undercutting their confidence or the quality of their work, but I knew there was more to the story than this clean paper they put out," she says.

Former scientists also have practical skills that come in handy, like a strong background in basic scientific concepts and research methods, and a deep grasp of their own field. Writers with scientific training can also decipher statistics and quickly digest technical papers that might take other journalists hours to slog through.

Beyond that, however, scientists and journalists share some fundamental traits. Shen cites qualities like curiosity, skepticism, and persistence, particularly in the face of unconvincing answers, which she says she encountered occasionally, for instance, while reporting policy pieces for *Nature*. "I think both scientists and journalists are tuned to push back on that," she says.

One Step at a Time

The territory between science and science writing can seem imposing and mysterious, but many scientists have navigated it successfully, each following a unique path. There's probably no wrong way, but there's no clear road map either. "Every person is an *n* of 1," Shen says.

At first, that fact gave me anxiety. But then I found an analogy that helped me cope with the overwhelming task of changing careers. I discovered it on the *ScienceGeekGirl* blog, where Stephanie Chasteen describes her own transition from physics into science communication:

> Bacteria—thermophilic or acidophilic bacteria, for example—do not "know" that the hot spot or acidic island is "over there." They have no overall map of their surroundings to direct their movement in a straight line toward what they seek. What they sense instead is a *local gradient*—a small change, right next to them. It's a little warmer *that way*. They move slightly. They feel it out again. Move. Feel. Move. And feel. The resulting path is a somewhat jagged, but non-random, path toward the thing that they love. And so is mine.

3 Do You Need a Science Degree to Be a Science Reporter?

Aneri Pattani

IN 1974, SCIENCE journalists from nine European countries met in Salzburg, Austria, to discuss the future of the field. It didn't take long for a major divide to appear. Britain and Switzerland saw science journalism as a graduate profession, wrote Martin Sherwood, science policy editor for *New Scientist* at the time. They foresaw a future in which nearly all science journalists would have degrees in science. But Austria and the US were moving toward a trend of science journalists having degrees in journalism.

More than 40 years later, that debate continues. Do you need a science degree to be a science journalist?

It's a question unique to the beat. Journalists covering crime or education are not typically expected to have a degree in those subjects. But science journalism is often considered a more technical and knowledge-heavy beat. And while some people come to it from other journalism beats, or with degrees in English or history, many also have a background in science: space writers who studied astronomy in college or graduate school, or health reporters who were doctors in a previous life.

Such scientific training can prove useful in science journalism, but for every benefit the years spent earning a science degree may bring—from gaining technical knowledge to understanding a scientist's daily life and avoiding common science interpretation errors—there's a potential drawback, too. When science reporters from varied backgrounds analyze their experiences, they can see the way these journeys have helped or harmed them.

Deep Subject-Area and Technical Knowledge

HERE'S HOW A SCIENCE DEGREE MIGHT WORK IN YOUR FAVOR. One thing that sets science journalism apart from other beats is the technical knowledge it often requires. From knowing how to read research papers to delving into medical procedures and understanding theories of physics, there's a lot to digest. Having some of this knowledge from a science background can make it easier.

Even just having a basic scientific vocabulary—understanding concepts such as statistical significance and peer review, and appreciating the difference between hypotheses and theories—can be useful, says Sarah Zhang, a staff writer at the *Atlantic* who studied neurobiology as an undergraduate at Harvard. In interviews for stories about genetic engineering, she's found it useful to show her understanding of basic molecular biology: the difference between RNA and DNA, for example. "That way when you're talking to scientists, they don't have to go through basic concepts or use imprecise language," Zhang says. "It also gives them a level of comfort that you're someone who has some technical fluency."

For freelance writer Laura Kiesel, one of the main assets of her graduate work in natural resources and environmental policy is how it taught her to read scientific articles and understand the nuances of findings. That recently came in handy, she says, in writing about the correlation between mass shootings and "toxic masculinity." She used many studies to support her case, so it was crucial to properly interpret the findings and understand how statistically significant they were. "I think that's an issue a lot of journalists and editors without science backgrounds struggle with," she says.

HERE ARE SOME POTENTIAL DRAWBACKS. Scientific expertise can have its downsides too, making it difficult to produce compelling stories for a general audience.

The first hurdle can come during the interviewing process. If a scientist feels a reporter understands the jargon of their field, they may be more tempted to use it during an interview. And jargon doesn't make for good quotes.

That's one reason why *STAT* reporter Rebecca Robbins, a former history major, feels her nonscience background is useful. There are often times in an interview where concepts go over her head, she says. "I think that demands some degree of humility to be able to acknowledge to the scientist that 'I don't understand what you're talking about.'"

But it ultimately benefits her readers, as she's able to explain difficult concepts in a way that's palatable for the average person. "I never, as a reporter, want to take for granted that people know what I'm talking about," Robbins says.

A deep background in a scientific subject can also sometimes make it difficult to write about that topic clearly and compellingly, says Carl Zimmer, an English major from Yale who has gone on to become the author of 13 nonfiction science books and a regular contributor to *STAT* and the *New York Times*.

It's a challenge he witnesses firsthand when teaching science-writing classes. Students with science majors are sometimes tempted to "dump data on a page," Zimmer says. But the craft of storytelling is just as important to science journalism as the science component.

That's why Zimmer insists that all science writers, especially if they're coming from a science background, should practice the craft of storytelling by reading great writing, analyzing what makes it effective, and using that to build their own narratives around the science they cover. "It's important to remember in science journalism that pretty much no one has to read what you write," he says. Most science writers, he notes, are not breaking news that affects people's daily lives. "So you have to ask yourself how you're going to keep them reading. And I've found that telling stories is a very effective way."

Being Part of the Science World

HERE'S HOW A SCIENCE DEGREE MIGHT WORK IN YOUR FAVOR. Studying science opens up doors to unique worlds: the day-to-day work inside a research lab, the behind-the-scenes of clinical care, the politics of grant funding. Being a part of these worlds can provide journalists from science backgrounds a special advantage in gaining credibility with sources, landing jobs, and understanding the daily lives of the people they interview.

Ivan Oransky, cofounder of *Retraction Watch*, says his medical background plays a large role in his journalistic work today. His MD helped him land some of his early jobs with trade publications, and later *Reuters Health* and *MedPage Today*. "In general, there is an appetite in a lot of newsrooms for expertise and people who know things but don't have to learn on the job," he says.

And it was his work writing for the now-defunct medical-student section of the *Journal of the American Medical Association* that eventually led him to launch *Retraction Watch*, which covers retractions in science journals. "It's really about the process of research at a meta level," Oransky says, "so you need to understand it to cover it." He picked up that knowledge, which includes understanding how misconduct can poison scientific research, from years of tracking the latest health studies in medical school.

Similarly, the time Zhang spent in a research lab studying fruit flies made her privy to the daily life of a scientist. She learned how research papers are written, what it means to be first versus last author on a report, and how to look for the funding sources behind a study. That insider perspective has helped her understand how to approach process-oriented stories and scientist profiles, she says. "Science is not just a bunch of facts to memorize, but a process of discovery," she says. "I try to capture that in my writing."

Holding a scientific or medical degree also affords writers some level of automatic credibility among scientists. Oransky says when he started out in journalism, he made sure to include "MD" in his email signature. "I knew it would make doctors comfortable talking to me," he says. "And they were probably more likely to call me back because of that."

Gaining credibility can be more difficult for reporters without a background in science. Zimmer says people are sometimes "shocked or mortified" that he wasn't trained as a scientist. Some readers who disagree with his work for ideological reasons will even try to use his English degree against him—by arguing that he doesn't have the necessary expertise to be trusted. "But that doesn't bother me," Zimmer says. "I can show scientists the articles I write and let them judge whether I got it right or not."

HERE ARE SOME POTENTIAL DRAWBACKS. Being part of the science world can also raise concerns about objectivity and a journalist's ability to see the bigger picture. Science journalists may be accused of bias when writing about their own field. And sometimes that criticism is valid, Oransky says. "The question is, are they able to distance themselves from people who [could have been] their colleagues? Do they get defensive when someone questions their former profession?" That can interfere with the journalistic goal of considering and fairly representing multiple perspectives.

Being deeply embedded in one profession can also make it difficult to step outside that world. Zhang says she spent much of her first year in journalism writing solely about neuroscience. "In some ways, it can be hard to get out of your little hole," she says. "But I think that goes away eventually."

That's where Zimmer sees his background as an advantage. It leads him to "tell stories that start from a different place." For instance, his latest book, *She Has Her Mother's Laugh*, is about heredity. Scientists he's talked with often assume that by heredity Zimmer means genetics. But he doesn't view heredity solely through a scientific lens; he thinks of the idea of what we carry on from our forebears as a powerful societal concept, as well. "I think that my interest in the concept as something that guides people's lives and that they use to define themselves comes out of my nonscience background," he says.

Approaching topics that way can engage a larger audience, Robbins adds. *STAT*'s audience, she notes, includes not only scientists and those in the medical industry but also readers from other backgrounds. Robbins wants her stories to appeal to all of them. "I would never say I'm trying to just write for people in the industry," she says. "Ideally, I'd like my stories to be as widely read as possible."

Understanding Consensus and Avoiding False Balance

HERE'S HOW A SCIENCE DEGREE MIGHT WORK IN YOUR FAVOR. Science journalists have the power to affect public perception and, thus, public policy. It's important that journalists have enough scientific knowledge to wield that power responsibly, Kiesel says. One essential skill is the ability to avoid setting up false equivalence.

This journalistic sin is especially prominent in climate journalism. For years, journalists portrayed the evidence for and against human-caused climate change as equal, despite an overwhelming consensus on one side. That contributed to a persistent and, many argue, damaging misunderstanding among the American public.

It was recognizing the lasting legacy of that error that motivated Kiesel to go back to school to study environmental policy, six years after earning her undergraduate degree in English and journalism. What too few journalists understand, she says, is that "science journalism can't be approached the same way an ideological topic is approached." Reporting on policy-relevant issues such as climate change is not about simply reporting both sides—it's about understanding what the scientific consensus is and where the evidence is strongest, and then conveying an accurate picture of that to the public. "I see a lot more intricacies in these topics than I did before I went to a science program," Kiesel says.

HERE ARE SOME POTENTIAL DRAWBACKS. While deep knowledge about science can help journalists avoid false balance and produce more thorough stories, the passion that often comes with that knowledge poses the danger of leading journalists into the realm of advocacy. That can raise concerns about objective reporting. After all, communicating the science you know to readers is not the same as reporting out a story to include diverse perspectives and provide more nuance, says Robbins of *STAT*. It is the latter, she notes, that helps makes stories richer.

It also makes a piece stronger to get a consensus—when one exists—from speaking with multiple sources rather than relying on your own knowledge, Robbins says. In a story she wrote about a new genetic fertility test, she conducted in-depth interviews with several reproductive endocrinologists before concluding that the company's claims about the test were oversold.

"The ability to report well, write well, and ask the right questions can help someone understand what they need to understand," she says, "even if you don't have a science degree."

Bottom Line: Choose Your Own Path

There is no scientific way to become a science journalist. No perfect formula. No universal law to depend on. Instead, it's about gaining the right combination of skills and knowledge, which can be done in a number of ways. Some, like Zimmer, read textbooks in their free time, while others, like Oransky, spent years training in medical school.

No matter which path one chooses, the most important thing is to simply be open to learning all the time, says Oransky. Studying biology, chemistry, or any other field of science can give you a foundation for science journalism, but ultimately this career is built on curiosity. "If you're curious and always willing to prove yourself wrong, then you're going to be a good reporter," he says. "As much as a degree helps you do that, it's great. But if it's going to close your mind and make you think you know everything, then journalism is not for you."

4 How to Break into English-Language Media as a Non-Native-English Speaker

Humberto Basilio

"HOW THE HECK did you do that?" one of my colleagues asked me soon after I published my first piece in the science magazine *Eos*. I couldn't believe it, either. Before I pitched *Eos*, the last time I wrote something in English had been during high school. "Good mentors and a bit of luck," I told her.

During the four years I studied journalism in Mexico, no professor told me it was possible to write for media outlets in other countries, much less to make a living from it. Of course, I never asked them about it, because I never imagined it could be possible.

Sending pitches to such big-name media outlets is intimidating even for journalists whose first language is English. It is not uncommon for non-native speakers to lack confidence in their English, their connections, and their understanding of a rather confusing world.

Still, many journalists who did not grow up in English-speaking countries have built successful careers working with major media outlets. They have built relationships with editors who see their backgrounds as a differentiator that gives them an advantage as writers.

Now, when I think about how I broke into English-speaking journalism, I think less about luck than about the practical advice and support I have received from other journalists. In the end, luck is a result of taking action.

Is Your English Good Enough?

While working as a staff reporter at the G1 news portal at *Globo* in Brazil, Brazilian journalist Mariana Lenharo—like me—never considered writing for an international publication. "I thought that with my level of English I wouldn't be able to do it," she says. Then, she realized that some of her more experienced colleagues, whom she knew had similar English-speaking and -writing skills to her own, were already doing so.

After six years of writing for publications like *Nature*, *Scientific American*, and *Undark*, Lenharo says that working in another language will always be difficult. "I still take longer to write in English compared to a native-English speaker, and I still make mistakes, but I believe it does get better with time."

So just how proficient do you have to be in English to start publishing stories? When I started, I was able to read texts and understand English speakers, but I worried that I didn't know enough higher-level vocabulary. And that's not the only challenge, of course—you also have to be comfortable writing and interviewing in English.

But you may not need to know as much English as you think. While writing my first story for *Eos*, I was very nervous because I thought my English vocabulary was not extensive enough. Then science journalist Emiliano Rodríguez Mega, who was one of my mentors at the time, gave me a piece of advice that helped me transform that insecurity: In journalism, writing in a simple way is necessary. "English sentences are usually much more concise and direct than in Spanish. Writing with simplicity is an advantage for you," he said. He meant that having a somewhat limited vocabulary can actually be an advantage: It forces you to write short, straightforward sentences, which can be helpful to the reader.

It's also important to remember that you don't have to sound like an English-speaking robot. *Hakai Magazine*'s editor-in-chief, Jude Isabella, says that editing non-native-English speakers is also an enriching language exercise for editors. "I love charming use of English words in a way that I would never use but isn't necessarily wrong," Isabella says. "I try to leave those [in] as much as I can because it's an unexpected way to use the language and I think that it gets readers interested."

There are also tools you can use to help polish your language and introduce you to words you don't know. Lenharo and other journalists, including me, often use a translator called DeepL. It manages dozens of languages and provides a variety of synonyms, which allows you to keep your word choice from being repetitive. (For one thing, it has helped me find more conjunctions and connectors—beyond *although*, *yet*, or *even so*—to string my clauses and paragraphs together.)

DeepL translations have a less artificial tone than Google Translate. For instance, I translated from Spanish to English a short explanation of what atmospheric aerosols are. Google translated it as, "Atmospheric aerosols are the set formed by breathable air and solid or liquid particles that remain in suspension." DeepL, on the other hand, says, "Atmospheric aerosols are the mixture of breathable air and solid or liquid particles held in suspension."

The difference is small, but crucial. Also, if you click on each section of the sentence, DeepL offers a variety of alternatives in case something doesn't sound convincing or natural. And although it has a range of paid subscription plans, the free version (like the free versions of other tools included in this story) works great.

Another helpful tool is Grammarly, which not only corrects basic grammatical errors but also offers syntax recommendations and generates a report on the clarity, correctness, and engagement of your text. Grammarly works when writing any text on your computer, not just in the Grammarly app. The free version is quite good, but you can also get a premium version.

Of course, with both Grammarly and DeepL, you have to check to make sure that the translation and tips make sense. It's also important to continue to challenge yourself to improve instead of leaning too much on the tools.

In the beginning, I would write complete paragraphs in Spanish, paste them into DeepL, and give the translation a final check to verify that nothing looked or sounded weird. Then I realized that writing directly in English, and using the tool for translating words that I didn't know or to find synonyms when I thought that I was overusing words, was the best option to push me to practice my writing.

It also helps to know where you are most likely to make a mistake. "I always make mistakes with prepositions like *in* and *on*," says Lenharo. "I know that I'm gonna get those wrong, so I always double-check." DeepL also helped me to stop misusing the apostrophe in possessive nouns, something that had always been my torment because apostrophes are rarely used in Spanish.

You should also consider the publication's writing style when pitching. Some places prefer writers to take a just-the-facts approach, while others want some literary flair. "Being aware of our English-writing skills will help us know which media we can adjust to," says Sebastián

Rodríguez, *Climate Home News*'s special projects editor. Understanding your abilities will help you know what you can propose and deliver to an editor, Rodríguez says.

One of the hardest parts of writing in English is interviewing. To make it easier, Lenharo recommends preparing English-language questions in a scripted format ahead of time. That way you can have a visual backup in case things get tough. (Of course, this can be helpful in any language, even if you're fluent!)

It is also very useful to have the complete interviews transcribed so that you can refer back to them whenever necessary. (Again, this is a good standard practice.) This is particularly useful with sources who use a lot of jargon or speak very fast. I first started transcribing everything manually because services like Trint or Otter were too expensive for me. But some months ago I started using Pinpoint from Google Journalist Studio, which transcribes in seven different languages and, unlike many other apps, is free.

Breaking In

Sarah Lewin Frasier, an editor at *Scientific American*, says that as long as writers can express themselves clearly and communicate the science accurately, they don't need to have native-English proficiency. "If the idea is strong enough, I'm sort of more motivated to go back and forth to develop the pitch with the writer and to work with them," she says. This is particularly true if you don't have clips of previously published work in English to demonstrate your writing skills. A well-structured pitch should show that the writer understands the subject they want to explain, and that they can identify a good fit for the magazine. "That will tell me a lot about how you would handle writing something, even if I can't see a sample," Frasier says.

To find a good fit, it is necessary to know the magazine or publication well. The most recognized media outlets typically have their own submissions section on their websites, where they explain in detail what kind of stories they are looking for, in what formats, and even how many words or paragraphs to include in a pitch. You can often find them by searching for the publication name and "pitch guide" or "submission guide." Generally, these outlets are not looking for completed stories.

Being a frequent reader of *Discover* and *National Geographic* allowed Malaysia-based journalist Yao-Hua Law to understand the types of stories those magazines told and the formats in which they presented them: long feature stories or brief articles based on a single scientific study.

In 2013, Law sent a 600-word story to *The Scientist*, and while sending a completed piece instead of a pitch worked twice for him in two months, thinking that was the way to work got him into trouble. "It misled me into thinking that this is an easy career and I will always get it," Law says, "but my next accepted piece was like two months later. [Then] I realized how difficult it was."

As a result of that experience, Law began researching and asking questions about how to write pitches properly and what were the most appropriate ways to approach editors.

According to Martin Enserink, *Science* magazine's international news editor, short stories like Law's are the best way to make the first approach to an editor. *Science* rarely assigns a full feature story to someone they haven't worked with before, Enserink says. "Pitch a [600–800 word] news story that we simply can't refuse because it's too interesting," he says. "In that way, you can see if you like working with us, and we see if we like working with you."

News agencies such as *Thomson Reuters Foundation News* or SciDev .Net, which tend to publish short news stories in an inverted pyramid format, can be some good options to start. *The Xylom* is a publication that is explicitly looking to help early-career journalists to publish their first pieces. Once you begin to get a better handle on English, you can take the next step and propose stories for media whose style is more narrative, which will let you begin to experiment with a more complex vocabulary.

While finding interesting scientific studies that staff writers haven't found and written about yet can be tricky, being a journalist based outside the US or the UK can also be an advantage. Many of the big press services like EurekAlert! don't include as many studies or papers from non-English-speaking countries. You can also keep an eye open for small local science stories, even if they do not come from academic papers.

Italian journalist Lou Del Bello, who lives in Delhi, India, used this tactic to pitch a short story in 2017 to *New Scientist*. Endemic ancient shrimps from a small lake tucked away in the mountains of central Italy were endangered by climate change and an earthquake that hit the area in 2016. She first heard about the shrimp from her parents, who live in Italy. "That's the trick," says Del Bello, speaking from her experience both as a writer and as an editor. "Try to find something that's unseen and take advantage of your unique language skills."

You can also use your expertise when a local story goes global. Lenharo's first time writing a story in English came in 2016, when she was a fellow with an International Center for Journalists program that placed foreign reporters in US media outlets to learn about digital tools. She was assigned to *Mother Jones* magazine and worked there for a month. That same year, the Zika epidemic broke out in Brazil. Although it was not the program's objective, the magazine's editor suggested she write an article.

Overwhelmed, she told her editor that she didn't think she could write in the publication's style. As they discussed it, though, Lenharo began to feel more confident because Zika was a topic she had already been covering in Brazil, and although her English writing wasn't fluent, she knew the subject well. "My editor told me that if I made any [writing] mistakes, he would correct them through the editing process. That made me feel much more comfortable." Lenharo wrote a story about the unusual ways Zika was affecting Brazilians, and received a boost of confidence after editors and writers at *Mother Jones* and other media outlets told her it was a great article.

That experience helped Lenharo gain more confidence in her abilities. "Objectively compare the stories you publish in your own country with those in the international media," she says. "You may realize that you already have what it takes to reach that goal."

Once you have an assignment, it is important to check with your editor about the publication's policies around things like when to disclose a conflict of interest, whether you can show your draft to sources, and so on, as this is different at every publication. Being clear up front about expectations will help you avoid potential problems down the road. You should also ask what the workflow around editing will be. Sometimes writers will work with more than one editor for a piece, which will make the editing process more extensive. (I'm writing these lines in the fifth round of editing after the first draft, and also have recontacted two of my sources at least three times to refine details.)

Fact-checking procedures can also be a surprise. When Lenharo worked with *Undark* for the first time, she was asked to highlight every fact she wrote and add a comment with the exact source that the information came from. In the case of interviews, she was asked to send audio files and transcripts pointing to the exact minute where her interviewees said what she was writing. Although it is a very intense process, Lenharo says, in the end "you feel much more confident [especially] with long stories." Such arduous processes are more common in long-form and investigative reporting. For many other stories, especially short ones, adding links to authoritative documentary sources that back up key statements may be enough.

Getting Used to Heavy Editing

Receiving an edited story with much of your prose crossed out in red can be discouraging, but it is a normal part of the process—and in the long run, writers can learn and improve a lot from it.

Enserink says that when he's working with writers who don't know the language quite as well, stories might need to go back and forth a few more times than expected. That's why editors need to have patience and plan for those stories to take a bit of extra time.

Line editing from a fluent English speaker will always make the stories shine, says journalist Dyna Rochmyaningsih, who lives in Deli Serdang, Indonesia. She says she readily accepts the language changes her editors make to her articles. The editing makes fixes to word use and structure that always takes the vocabulary to the next level. "I think most native speakers will understand that English is not our first language, and they will be happy to edit," says Rochmyaningsih.

Sometimes, editors will have to make major changes to stories to make them fit the allotted space or make them suitable for their audiences, especially if they are short articles where words are limited.

However, writers shouldn't feel obligated to accept every edit from an editor, particularly when they understand the local context best. Enserink says it's very important for writers to take an active role in the editing process. "Writers from other countries know better [their countries'] cultural issues," he says. "It's your story. If there's something that you didn't like in the edit, tell me—don't just accept all changes."

In 2018, Rochmyaningsih reported how the measles immunization rate in Sumatra was falling because some Muslim families were refusing to inoculate their children with vaccines that contained pork elements such as gelatin and the enzyme trypsin. Tired of seeing the mainstream media only publishing articles about how vaccines had been "forbidden" by Islam, Rochmyaningsih decided to write an in-depth story that explained certain Islamic legal terms and described the situation in greater detail, recognizing that some Muslim clerics took a more nuanced view of the vaccine-production process and permitted use of vaccines that contain pork products. During the editing process, some details that Rochmyaningsih considered important were edited out. But after she communicated her point of view, her editor understood and reinstated the deleted material.

Building a Community to Rely On

The journey will always be more bearable with colleagues by our side to accompany us. In 2018, while Lenharo was a student at Columbia University Journalism School, a colleague invited her to join Study Hall, a digital publication and online community for freelance media workers dedicated to sharing job opportunities; organizing talks, seminars, and workshops; and more.

Through reading and listening to other colleagues' advice and experiences, Lenharo got editors' email addresses, sent her first pitches, and learned that having pitches rejected was much more common than she thought. She also learned that some big media outlets pay $1 per word or more (others, including many smaller outlets and those with less-healthy freelance budgets, pay significantly less). Knowing this helped Lenharo make decisions when negotiating rates.

Networking with other journalists is essential to staying updated about opportunities within any profession. "Build a community," says India-based journalist Disha Shetty. That includes following and DMing journalists who are doing what you aspire to do; joining WhatsApp and email groups to find mentors; or joining formal journalist associations. Shetty says those are all useful ways to get advice and find out about grants, fellowships, and opportunities to grow.

Advice for Editors Working with Non-Native-English Speakers

Collaboration between media in the English-speaking world and journalists outside these regions is mutually beneficial. But success requires effort on both sides.

Hakai Magazine's Jude Isabella says that having video meetings to get to know each other after sending the first draft is a practice that can improve communication and create a bond of trust and mutual respect. Of course, this will depend on whether the writer feels comfortable with their English-speaking skills.

Editors should also be careful to explain changes clearly. While explanations about why a sentence was rephrased or a paragraph rearranged are important for all writers, it's particularly valuable when it comes to working with writers who are using a second (or third, or fourth) language.

It's always important for editors to be empathetic, including when rejecting pitches, and even more so when working with writers whose first language isn't English, says Alex Ip, editor-in-chief of *The Xylom*. Taking a few extra minutes to give feedback to writers on why their pitches weren't accepted or pointing them toward other outlets where their ideas can fit better, can truly make a difference to a freelancer's career, he says. "You shouldn't just be the person that says 'yes' or 'no.'"

Ip also suggests that just as many journalists track the diversity of the sources they've included in their stories, editors should regularly ask themselves how many writers who don't use English as their first language they've commissioned. Then, actively look for new writers and keep an open mind to new ideas. "I believe that everybody can write a good story if we give them a chance to do it and the suitable resources, trust, time, and patience."

In my case, joining the Mexican Network of Science Journalists (RedMPC) was a big step. There, I met the mentors who would later help me gain the confidence to send my first pitches. Similarly, other countries have their own national science journalism associations. Joining or even just exploring if your country has a national association in the membership list of the World Federation of Science Journalists (WFSJ) might be a good starting point for finding other professionals. And even if your country doesn't, other organizations, like the National Association of Science Writers, the Society of Environmental Journalists, and the Association of Health Care Journalists accept members from around the world to help them get immersed in the craft.

"Connections are the most important thing for a lonely freelancer," says Rochmyaningsih, who through WFSJ found a journalism training space, her first mentor, and a network of contacts that would help boost her career. "Networks with journalists generate networks with scientists, and that means you have more chances to write more science stories," she says.

And once you have colleagues to lean on, it will be easier to leap into the void.

5 Feeling Like a Fraud: The Impostor Phenomenon in Science Writing

Sandeep Ravindran

SITTING AT MY desk at my first science-writing job, I couldn't help thinking it was only a matter of time before my deception was uncovered. Any second, I thought, my editor would walk in, tap me on the shoulder, and tell me she'd made a mistake in hiring me. Objectively, I knew I was doing fine—I was turning in my work on time, and my editor was happy with what I turned in. But I just couldn't shake the feeling of being a fraud, of somehow having fooled everyone around me into believing I was a competent professional when I clearly was not.

It turns out I'm far from alone in feeling that way. A lot of people have trouble owning their accomplishments and worry they're not qualified for their jobs despite evidence to the contrary, says Frederik Anseel, a professor of work psychology and behavioral economics at Ghent University in Belgium. This so-called *impostor phenomenon* was first described among high-achieving women, in a 1978 paper by psychologists Pauline Clance and Suzanne Imes. It's since been shown to affect both men and women, and is commonly known as the "impostor syndrome" (though it's not a true psychological syndrome in the clinical sense).

Researchers have estimated that 70 percent of the general population has experienced the impostor phenomenon at some point, and it's a concept that seems to resonate with many.

"I've had people tell me, 'You've just described to me something that I have felt for so many years but I didn't know there was a term for it,'" says Kevin Cokley, a psychologist at the University of Texas at Austin. His work on the impostor phenomenon focuses on how it relates to academic and mental health outcomes among ethnic minority students. Several studies, Cokley says, have shown that "people who have higher feelings of impostorism are more prone to having symptoms of depression and anxiety." In the worst cases, impostorism can become truly debilitating, and those affected may benefit from professional counseling.

Impostor syndrome has gotten more attention in the past few years, with a number of famous figures—from Facebook executive Sheryl Sandberg to celebrated writer Neil Gaiman—publicly sharing their own experiences. That increased attention may be helpful. Often, Cokley says, "people who are experiencing or struggling with impostor feelings struggle alone. They think that they're the only ones feeling that way."

Most research on the impostor phenomenon has focused on graduate students or medical students, but it's been identified in people across a wide variety of careers. Impostorism seems especially common in competitive and creative fields, and those where evaluations are subjective.

Journalism certainly fits the bill. The feeling of being a fraud is also common in fast-changing fields such as technology or medicine. "So if you also happen to be a journalist in technology, in science, in medicine, there's also that sense of not being able to keep up like you should be," says Valerie Young, author of *The Secret Thoughts of Successful Women: Why Capable People Suffer from the Impostor Syndrome and How to Thrive in Spite of It.*

Anecdotally, the impostor phenomenon seems to be common in science journalism. Many of the journalists I talked to for this article had either experienced it or heard of colleagues experiencing it. A session on impostor syndrome at the 2013 ScienceOnline conference generated much discussion among journalists on Twitter. [The articles in this book were written before Twitter was renamed X.]

"It certainly rings true for me," says Laura Helmuth, who says she felt like an impostor when she recently started a job as the health and science editor at the *Washington Post*. "I feel completely incompetent, I just hope nobody notices," says Helmuth.

It's particularly common to feel like an impostor early in one's career. "This is something that does get better over time," says Helmuth. But it's situational, and can crop up when starting a new job or taking on a leadership position. "The impostor syndrome never really goes away, and each time you do something new, of course you have to deal with it again," says Helmuth.

For most people, the impostor phenomenon is a normal part of developing a professional identity, says Holly Hutchins, a professor of human resource development at the University of Houston. But at the more extreme end of the impostorism spectrum, the experience can have tangible effects on mental health, job performance, and career decisions. "When it's persistent and you begin having effects of depression or anxiety and it's prolonged, that's when it becomes an issue," says Hutchins.

Overblown Concern or Hidden Scourge?

For those who've experienced it, it can be hard to imagine not feeling like an impostor at least sometimes. But not everyone suffers from this problem. Several science journalists I contacted said they had neither experienced feelings of impostorism nor heard of any of their colleagues experiencing it. Some were skeptical about whether it was a real thing or whether it was worthy of discussion.

"I haven't experienced it personally and I don't really recall observing it in anyone that I've mentored or edited," says freelance writer and editor Robin Lloyd, who is also an adjunct professor at New York University. "I really question whether the impostor phenomenon is a useful concept," she says. She believes it's more important to help journalists deal with their normal feelings of insecurity and self-doubt, and says she does so by calling out good work when she sees it. "Instead of reifying this questionable concept, instead of pathologizing emotions that are quite common in any self-reflecting human, I think we need to do a better job of letting people know when their work is outstanding or shows particular strengths."

Sociologist Jessica Collett of the University of Notre Dame ran into similar sentiments while researching the impostor phenomenon. "It was very interesting to sit down with people who were like, 'I have no idea what you're talking about,'" Collett says. "For people who have felt that impostorism, it's so palpable." The impostor phenomenon is, in fact, correlated with general feelings of self-doubt and low self-confidence, she says. But studies have shown that despite its overlap with feelings of anxiety, low self-esteem, or depression, impostorism seems to be its own thing.

It's still unclear what exactly causes the impostor phenomenon, although some studies have found that it's associated with aspects of family upbringing, such as having parents who are overprotective or have high academic expectations; personal factors, such as a history of anxiety or depression; and the presence of certain personality traits, such as perfectionism. People who experience the impostor phenomenon tend to have trouble taking credit for success, often attributing achievements to external factors, such as luck or timing. They also tend to beat themselves up about their failures, blaming their own lack of competence.

The impostor phenomenon can be hard to spot in others, because ironically, those who experience it generally do very well at their jobs. "It results in you feeling like you have to work even harder to prove yourself, and that can look like higher levels of achievement," says Cokley.

That rings true to freelance journalist Nadia Drake, who says she considers every assignment a test to be passed. "Every story has to be the best thing I can do, because I don't want some editor to discover that, 'Whoops, Nadia Drake actually isn't that great,'" she says. Every time she takes on an assignment, she thinks to herself, "Don't let this be the one that exposes how inept you are," Drake says. "The consequence of that is that I just work really hard all the time, and I'm very nitpicky about things that I probably don't need to be spending so much energy or effort on."

Feeling Like a Fraud Takes a Toll

But even when you're doing well at your job, the constant effort to keep others from finding out you're a fraud can take its toll. "[For] those people who score very high," says Collett, "the chances are good that impostorism is debilitating to them. It is in some way negatively affecting either their mental health or their tangible productivity." Studies show that the impostor phenomenon is highly associated with emotional exhaustion, which can lead to burnout. "You might get higher grades, or perform even better at your job, but at what cost?" says Cokley.

Professional success may not serve as an antidote, either. In fact, it can actually exacerbate the impostor phenomenon by making people feel more exposed and more likely to be found out. That can lead to a fear of success, which causes people to actively avoid putting themselves in a position to succeed. "There are people for whom it is so debilitating that they just never try," says Young. "They just fly under the radar, because it's just safer there."

A recent study found that experiencing the impostor phenomenon can decrease people's career planning and their motivation to pursue career goals or leadership positions. Other research, not yet published, found that among women in academia, increased impostorism played a major role in their greater propensity to "downshift" their professional goals and to give up on ambitions of becoming tenure-track research professors, compared to their male counterparts.

What does this look like among journalists? For one thing, it may manifest as a reluctance to apply for awards or pitch high-profile publications. "I'm always talking myself out of pitching things," says freelance journalist Diana Crow. According to Drake, "I spent a lot of time not applying to awards for a couple of years." When she did apply, she won some awards, which of course brought its own impostor feelings. "There's a little bit of wondering whether what won an award is actually award-worthy," she says.

Could the impostor phenomenon play a role in gender differences in pitching? As a woman, "you're less likely to pitch, and if you do pitch and you get rejected, you're less likely to tweak [the pitch] and try again," says Young. She suggests that this difference may partly explain why men account for more op-eds and science features in high-profile publications than women, although Young suggest that implicit bias may also play a role.

The Impostor Phenomenon in Women and Minorities

Although the impostor phenomenon was initially identified in women, a lot of subsequent research has found no gender differences in its prevalence. That's not to say that its effects are the same in women and men. For example, impostorism may play a more important role in academic achievement in women than men, Cokley found. "For women, higher impostor feelings were linked to higher grades, but we did not find that in men," he says.

When it comes to dealing with feelings of impostorism, women often seem to get the short end of the stick, whether that manifests in fewer pitches or "downshifted" career goals. "Even though there are many men who feel like impostors, it holds men back less because they're gen-

How Do You Score?

Researchers use a tool called the Clance Impostor Phenomenon Scale to measure the severity of a person's sense of being an impostor and to differentiate those feelings from anxiety and low self-esteem. (To find out how you score, visit bit.ly/345iHsn.)

erally better at compartmentalizing," says Young. "They're more apt to say, 'Well, I'm just not going to think about it.'"

There may be parallels to research on how men and women respond differently to burnout and stress, says Anseel. "Women will seek more social support and talk about things, while men will try to tackle the problem on their own or will just shut up about it and hope that it will pass." he says.

Women do appear to be more open about their feelings of impostorism. "Some of the research has suggested that the difference between men and women is that women are just much more likely to talk about it, and also lean on social support," says Hutchins. In contrast, she says the men she interviewed for her research told her that "[impostorism] would be seen as a weakness, there's no way they would talk about it with their male colleagues." In the small sample of science journalists I contacted—about 20 people—many more women than men said they had experienced the impostor phenomenon or heard about it from colleagues, and they were also far more willing to talk about it publicly.

Some researchers believe the impostor phenomenon may have an especially strong effect on any underrepresented minorities, particularly among groups for which stereotypes about competence are prevalent. Few studies have examined the impostor phenomenon among underrepresented groups other than women, though.

In a study of the impostor phenomenon among ethnic minority college students, Cokley found that both impostor feelings and the stress of being a minority student were associated with psychological distress. What's more, these students' impostor feelings were more strongly related to their mental health outcomes than was the stress they attributed to being a minority. Those findings hint that the lack of diversity in newsrooms might play a role in increasing minority journalists' feelings of impostorism. "As a minority, you feel that," says freelance journalist Sujata Gupta. "If you think about a newsroom or magazine culture, I've always felt like one who doesn't quite belong."

How Do I Get Over It?

I know what you're saying: "Yes, it's ridiculous that Neil Gaiman or Sheryl Sandberg or Laura Helmuth ever feel like impostors. But that doesn't mean other people aren't faking it. How do I know I'm not the only true impostor around?"

"It's a false question," says Young. "Somebody may not be the best qualified for a job, but it doesn't make them an impostor. Impostorism is an internal experience, of internally not being able to own your accomplishments," she says. If you really need to convince yourself of your competency, it may help to make a list of all the positive things that you've achieved on a weekly or even daily basis. "People who experience impostorism tend to minimize all the evidence that challenges their feelings of impostorism," says Cokley. "So sometimes you have to be purposeful and intentional in getting them to recognize the good things that they've actually done."

Multiple journalists I interviewed declined to discuss the subject on the record because they didn't want to publicly admit a lack of confidence in their abilities. That's too bad, suggests Helmuth, "because as soon as you start talking about it, that kind of demystifies it."

Young agrees. "I do think it's important to name it, to give voice to the feeling, because there's so much shame attached, and you think you're the only one," she says.

There's been relatively little research done on interventions to deal with impostorism, but studies suggest that social support can help. "We see that social support is sort of a buffer against the negative effects, and if you're a freelancer working on your own, this could strengthen feelings of inadequacy," says Anseel.

Social Support Is Key

That's where belonging to a freelance group that functions like a true community can help. "I've had a really great support network, and it can make such a difference, because if you start to feel this kind of thing, you can sort of nip it in the bud," says Christie Aschwanden, lead science writer for *FiveThirtyEight* and a member of several small, informal groups of science writers who discuss the trade, offer one another support and advice, celebrate successes, and commiserate about disappointments.

Writers can also be overly self-critical because they see every flawed draft of their own articles, and these suffer in comparison to everyone else's polished, published pieces. Becoming an editor can change one's perspective and help with feelings of impostorism, says Helmuth. "You realize, even with the very best writers, nobody writes it perfectly the first time." She also points to *The Open Notebook*'s Pitch Database as a way to get a behind-the-scenes look at pitches and [corresponding stories].

Finding sources of honest and constructive feedback can also help combat feelings of being a fraud. Ideally, a writer's editors would provide such feedback—but that doesn't always happen. "Some editors can take for granted the things that are working and just focus on the problems, and I think that can be a problem for somebody wrestling with impostor syndrome," says Helmuth.

That's why it's worthwhile to seek out other colleagues and mentors who can help writers make sense of their success or failure and set realistic expectations for themselves, says Hutchins, who has found that mentoring can help people deal with impostorism. "Good mentors can create a safe space for individuals to talk about some of these concerns and struggles. Often that's half the battle," she says.

Feelings of impostorism can unfortunately make it more difficult for people to reach out to mentors, particularly for those who already feel like they don't belong. "I don't know how to find somebody that I would really connect to," says Gupta. "The thought of reaching out to somebody who is seemingly so far above me is very unnerving," she says. Helmuth agrees that this can be a problem, particularly for those

What about Real Impostors?

Some people have a Bizarro version of impostor syndrome. People who manifest what is known as the Dunning-Kruger effect—named for the psychologists who identified the phenomenon—are overconfident despite being incompetent. They're too clueless to realize how clueless they are, which leads them to vastly overestimate their abilities.

who are new in the field. "One thing I would want people to know is that people like helping, and people like giving advice," she says.

Competent People Who Don't Feel like Impostors: What's Their Secret?

Some people have suggested that impostor feelings help a person stay humble, and wonder whether those who never feel like impostors are likely to be overconfident or arrogant. "I think that kind of assumes that you're *either* arrogant *or* you feel like an impostor, and that's kind of a false choice," says Young. "You can still have humility." Not feeling like an impostor doesn't automatically mean someone is either arrogant or is clueless about their own incompetence, she says. "Some of it is they have a good, healthy sense of their limitations."

At some point it all comes down to changing how you think about failure and competence—or what psychologists refer to as *reframing*, says Young. "The only difference between people who think they're impostors and people who don't is that in the same situation that would evoke impostor feelings, they think different thoughts," she says.

The journalists I spoke to who hadn't experienced the impostor phenomenon certainly seemed to think about success, failure, and rejection differently than those who experienced impostorism. "All the jobs that I've had, I felt that I deserved them; all the promotions that I've gotten, I felt that I earned them," says Lloyd. "When rejection or failure crops up it's disturbing, and you have to think it through and analyze it, but ultimately I tend to just get back to work and move on," she says.

Award-winning science journalist Alexandra Witze describes a similar way of countering self-doubt: "I just remind myself, 'You know what you're doing, you're good at what you do. Just do it.' I tell myself, nobody's going to push my career forward but me, so if I'm not doing it, it's not going to happen."

Such reframing is something freelancer Crow has been working on. She tells herself that "this is all a process of learning and practicing," and as she gains more experience, "I'll be able to blow my current self out of the water."

Of course, it's not always easy to change how you think, or to believe your new thoughts. "You won't believe it—that's why you feel like an impostor—but you have to keep going regardless of how you feel," says Young. "Feelings are the last to change." In other words: Fake it till you make it.

Organizations, institutions, and managers should also take the impostor phenomenon more seriously and appreciate its connection to outcomes such as job performance, work satisfaction, and emotional exhaustion, says Hutchins. "By not recognizing this, not including it in orientation, not training your mentors to recognize this and offer support, you're running the risk of losing your top talent," she says.

The best bet for an individual beset with feelings of being an impostor might be to try to change his or her mindset and forge ahead regardless. One could even view impostorism as an occupational hazard of working in a challenging field. "The trick," Helmuth says, "is finding the excitement of something that's confusing, something that you're not good at, and learning to enjoy that as an opportunity for learning new stuff. Which is easy to say and hard to do."

6 What Is Science Journalism Worth?

Kendall Powell

THE MONEY IN this job sucks. Let's get that out up front. Even from my first days of grad school I knew that a majority of my college classmates working in the corporate and finance worlds were making scads more money than I was—and probably ever would. But, as I smugly told myself, I love my job.

And I still do most days.

However.

After 11 years of freelancing, I've added a mortgage and two kids to the mix. As science journalism tosses about in the current (everlasting?) economic maelstrom that is publishing, it gets harder and harder to sustain myself financially. And I'm not the only one.

The combination of the collapse of print advertising, declining magazine subscriptions, the enormous availability of free content online, and the rise of the gig economy, has led to pay rates that have stagnated for the last 30 years, says Robin Marantz Henig, freelance writer and 2014–2016 president of the National Association of Science Writers (NASW). Freelance rates for science writing have not increased in the last three decades, and yet in 2014 it took $227.29 to purchase what cost $100 in 1984. This malaise has writers of all kinds talking and writing about their job security fears and inability to make a living wage in journalism.

The current economic state of science journalism is making it increasingly untenable to do serious and thoughtful work. Low and even declining pay rates are also pushing many of us to become what we've always feared—the harried reporter who turns in subpar copy that rests on thin reporting and shaky fact-checking because we literally can't afford to spend one more minute on it. It's also keeping the gates closed to more diverse perspectives from those who are less privileged—those would-be writers who must choose a career track with a real paycheck over a writing internship or a newsroom job.

"There is the disappearance of the idea that you can live the life of the mind and actually make a living at that," says Henig. "None of us got into this profession for the money, but we did expect to make a living wage."

It's good that writers are talking about the problem with each other, in the halls at conferences, in online private forums, and on blogs. While much of that may be intended as venting, sharing knowledge also increases our collective power. We need to continue to shine a light on this issue. Barring a miraculous influx of cash to the system, this is the reality that freelance science writers and editors must work within. The challenge for all of us will be finding ways to continue to produce quality work under these financial constraints.

Quality Suffers

"Both editors and freelancers are caught in the gigantic economic mess," says Dan Vergano, senior writer-editor at *National Geographic*. "But we're not willing to talk about economics enough in our business. We're caught in a big economic moment, and yet people continue to pretend that it isn't what's driving everything."

Although he's only been assigning and editing stories for a year, Vergano says he notices major differences in how articles are carried out compared with when he worked as a freelancer 20 years ago. "The pitches I get are much less well-developed, in general," he says, lacking the five or six meaty paragraphs that lay out the lede and nut grafs, a proposed structure, potential sources, a timeline for completion, and expected compensation. "In the gig economy, nobody has time for that," he says. "Writers gotta get a sense right away if you are going to bite or not, so they send one-sentence descriptions."

Such hurried pitching inevitably leads to editorial headaches. "Commissioning a piece [based on weak pitching] is not fair to the writer or you—especially if it comes in and it's not what you expected."

Writers' time for reporting and writing has also been compressed. Features that used to be written over the course of, say, four weeks are now given perhaps eight days. Writers are constantly calculating just how much—or how little—time they can spend on a story, often juggling multiple deadlines in a week. One area where that time compression becomes visible, Vergano says, is news stories filed in what he sees as a blog format—without proper sourcing, clear ledes, nut grafs, or structure, and sometimes in first-person voice.

It's not just the quality of initial drafts that suffers. The same forces that have suppressed freelance rates have also slashed the size of the editorial staff at most publications, while also increasing the amount of time they needed on each story. Rosie Mestel, chief magazine editor at *Nature* and former science and health editor at the *Los Angeles Times*, remembers that when her freelance budget at the newspaper was slashed, it became "tricky, as an editor paying crappy rates, to get really good writers." Although she assigned stories strategically—choosing writers she knew would want a certain story, selecting articles that were contained and easily executed, or offering a piece of ownership, such as a weekly column—those strategies only partly compensated for the paper's low pay. Some writers, Mestel says, were (rightly) businesslike with their time. "You could tell when the copy came in, and as editors we had to spend more time honing and polishing." But it was a trade-off Mestel was willing to make for writers whose reporting she trusted.

But as Vergano notes, lack of time on writers' and editors' parts means relationship-building and mentoring fall off, too. "The chance to coach people is so much less," Vergano notes. All too often, "we fillet their copy and publish it, and don't even explain [the edits]," he says. "And they don't get a chance to improve."

Apoorva Mandavilli, editor-in-chief of *Spectrum*, a foundation-backed journalism website that reports on autism research, says the gig economy poses an additional problem for her. The current economic atmosphere, she says, creates a tension for writers who feel that career success depends on publishing shorter pieces at the online versions of marquee publications, which usually pay less than their print counterparts. "I'm not at one of those high-profile, prestige publications," Mandavilli says. "It's frustrating for me that we pay well—$1.50 per word—but I can't always get the good writers to turn away from the high-profile bylines. They have to stay in the Twitterstream."

This tension keeps all of us, myself included, beholden to unbelievably low rates when we want a flashy byline. And it allows those publications' low offerings to suppress rates across the entire field.

Pennies for Your Words

It's not just magazine articles (both print and online) that suffer as a result of miserly pay rates. Books are equally endangered. Florida science writer Mark Schrope, a 15-year veteran, was offered a book advance that amounted to about one-fifth of what he calculated he could afford, for a book that would require a year of full-time work and extensive travel. In a recent discussion among the science writers of the SciLance community, Schrope asked, "Are there that many full-time writers who would go for this kind of deal?"

Others in the group noted that many writers might take such a deal if they can spread the work out and supplement it heavily with other writing work, if such a deal could be seen as an "opportunity cost" that leads to sharpened expertise and more work down the line, or if they are writing books for reasons other than the up-front money.

Emma Marris, whose first book, *Rambunctious Garden*, was published in 2011, and who is working on a second, shared her own reasons for writing books even though doing so is financially precarious at best: following her passion, being taken more seriously by editors, and wanting to land speaking gigs as another form of income. She also hoped that if her first book did well, she might earn higher advances on future books.

As a primary breadwinner with the first of three children about to head to college, Schrope couldn't take such an income hit, he told the SciLance group. So for the moment, his successful book proposal sits on the shelf. In the past, he said, "I've taken huge financial risks to become a writer and to take on big projects. But now's not the time."

Marris punctuated the listserv discussion with a comment on the struggle we all live in our writing careers: "It is a shame that something that is so widely appreciated by society is so very badly paid. Some of our books really change people's opinions and touch people's lives. And yet, [writing books] can only be done by those who are either spousally subsidized, bad at math, or just very stubborn."

While that does accurately describe me and many of my colleagues, there is more to this argument than "starving artist" lamentations. There is a greater good behind pushing for higher rates—namely, achieving an end product that serves the public well with more in-depth, well-reported, high-quality science journalism. Just as high-risk, high-payoff scientific research requires heftier investments of cash, so too does the kind of influential or investigative journalism many science journalists would like to practice.

Knowledge Is Power

So how can full-time freelance science writers pull our collective earnings out of the toilet?

The first step is getting some high-quality data on what science writers' words are truly worth. Luckily, that's already been partially tackled by the NASW Freelance Committee, which surveyed NASW members about their compensation in 2013. (Disclosure: I am cochair of the committee, and Siri Carpenter and Jeanne Erdmann, cofounders of *The Open Notebook*, are the previous cochairs.)

The committee hired a scientifically trained surveyor, Gary Heebner of Cell Associates, to help design a survey that would return the kind of granular data many freelancers thirst for: What are the going rates for different types of science-writing assignments? We wanted breakdowns by the word, by the hour, and by gender. [Editor's note: NASW conducted a similar survey in 2022, this time surveying only freelancers; the results with respect to pay rates were not meaningfully different from those described below.]

The survey yielded responses from 618 currently employed NASW members, 55 percent of whom identified as freelance writers or editors and 42 percent of whom said they earned their income solely from freelancing. Two-thirds of the respondents had worked in science writing for 10 years or more. For each respondent, the survey only collected pay rate data about the top five types of assignments that made up the bulk of their income. In other words, pay rate numbers by and large came not from hobbyists or dabblers, but from journalists whose majority income flowed from freelance assignments.

The survey results (available online to NASW members) provide a useful starting point for any science writer considering taking on a freelance assignment. For example, the 53 freelance writers who reported working regularly for the news sections of scientific journals reported that the median rate at such publications was $1.20 per word [rate in 2022 survey: $1.50 per word], and the range was $0.50–$2.00 per word [rates in 2022 survey: $0.50–$5.00 per word].

Based on these data, I draw a hard line in the sand for myself. I don't work for less than the median rate within a given marketplace because I feel doing so isn't good for me or anyone else. Not only does doing so make it harder to claw your way up to better rates over time, but it also drags down the whole field's pay rates. It can be easy to forget, but transforming complex concepts, discoveries, and scientific debates into engaging, readable, and accurate stories is a rare and valuable skill. (Remember, too, that freelance writers are a steal for publishers, who pay nothing toward the writers' insurance, retirement benefits, self-employment taxes, IT support, equipment, or overhead on home office spaces.)

The median rates and ranges from the NASW survey are the best information our field has for what science writing is worth in today's market. The best way to maintain and improve our pay rates as a field is to encourage all freelance writers to demand fair compensation for their work. Most experienced freelancers aim for a rock-bottom rate of $1.00 per word for magazine work and $0.50 per word for online or newspaper copy. Every freelancer is free to set their own rates, of course. You should decide where your line in the sand is, too.

Commanding More

Now that we have some data, the second step in freelancers' improving our lot is recognizing that the responsibility for doing so rests squarely on our own shoulders. The NASW survey rate tables offer writers a defined starting point for negotiation—something that should happen with each new assignment. Do the simple math. When considering a project, a writer should estimate how many hours it is likely to take and calculate the per-hour rate. For some projects, $2.00 per word translates to little more than minimum wage. For others, $0.50 per word can be a fast, easy paycheck.

When warranted, always ask for more. In 11 years of freelancing for more than 24 publications and 46 editors, I have been given a pay-rate raise without my asking exactly one time. (Thanks A.W.!) I have learned to negotiate for more money up front. It is the only way freelancers can earn merit-based or cost-of-living raises. No one else will do it for you. It may even improve the editor's opinion of you.

"Believe in your own worth and fight for it. If you don't ask, you don't get," Mandavilli says. The worst scenario is that the editor says it's not possible. And, as long as the request is made politely, it's never a reason to not work with a writer, she says. "It's absolutely okay, professional, and healthy to ask."

Mestel agrees, but she cautions writers not to overdo it—editors do curb the words assigned to writers who demand rates higher than the publication's standard.

If the client does agree to increase your fee, it's a personal stride that also benefits the entire community. Writers should also renegotiate for more compensation when asked to do a rush job or whenever an assignment requires more labor than the original agreement.

Finally, do your due diligence. Ask other writers to find out about pay rates and other contract terms at specific publications. Read the contract carefully. It should include a kill-fee clause with a minimum of a 25 percent payment of the fee if the article is deemed unacceptable. (Many of my own contracts have a 50 percent kill fee.) If an article is in acceptable condition and meets the original assignment, but the publisher decides not to run it for internal reasons, then the writer should be paid in full.

The contract should also clearly state the payment terms—ideally, payment upon acceptance (with "acceptance" defined) and payment made within 30 days. If either the conditions under which articles can be killed or the payment terms are not spelled out, consider asking to add such language to the agreement. NASW members can access sample contracts at NASW's The Fine Print database (members only).

If a writer uses every resource at her disposal to negotiate the best possible compensation and terms up front, then she can give that assignment the attention it requires. After all, at the point you sign on the dotted line, you are agreeing to deliver a quality piece of writing within the allotted time. It's a writer's responsibility to know what is expected to go into the making of that piece—how many revisions will be expected and what annotation, fact-checking, or art-gathering duties it entails—and to have factored that into the negotiations.

"Realize the situation you are in and force yourself to do good work anyway," Vergano says. "You will become valuable." (Try not to roll your eyes—he knows what he's asking isn't easy or even fair.)

He says one thing has not changed from the newsroom of 20 years ago. Writers who deliver good work on time and who are responsive to edits will be recognized by editors as "go-to" people when other projects, long-term contracts, or even the rare new staff position comes up.

"If you can be strategic in your thinking about how you do this, it will give you the chance to do quality work," he says. That might mean having a conversation with an editor about what you need to make the relationship work. Or maybe it means diversifying your client list so that higher-paying, shorter gigs can subsidize longer, deeper passion projects. "I think balancing your portfolio is number one," says Mandavilli, who also freelances. Her organization is one of a rare breed that still offers retainers—contracts that give writers a guaranteed base income, which can also help offset the time spent on passion projects.

"The world has shifted," says Mestel. Her advice to freelancers is to write more short, newsy pieces and fewer labor-of-love, longer pieces. "It makes me sad to say that. I don't think it would be that satisfying, nor would it serve the public as well."

Mandavilli also says more writers should look more closely at niche publications—often backed by foundations, scientific societies, or patient-advocacy groups that remain hands-off editorially—because they can afford to pay better rates. "There are places that still pay well and do journalism, it's just a different kind of journalism than what the *Atlantic* or the *New Yorker* is publishing."

While the median rates in the NASW survey tables are pretty depressing, all is definitely not gloom and doom. More than half of full-time freelance writers reported making $50,000 a year or more, and 29 percent made $75,000 or more. The survey found no significant differences between the pay rates garnered by men and women. (One limitation of the NASW compensation survey is that it did not collect data on respondents' race or ethnicity.)

And while the era of online uncertainty continues, it's also a fertile ground for innovation, in both the presentation and the funding of journalism. In the past year, for example, some freelance science writers have formed cooperatives and joined new crowd-funding platforms such as Beacon to bring their work directly to readers, who receive subscriptions in exchange for their support. David Wolman curated 15 past articles and book excerpts into an online collection for subscribers using the Creativist platform. Marris has used a campaign on Beacon to get her new book project on wolves off the ground. Another Beacon campaign, called Flux, was formed by six freelance environmental journalists who wanted to write stories about how the world is already bracing for the impact of climate change.

For me, these subscriptions are worthwhile: They deliver high-quality science news from trusted journalists right to my inbox. And other writers are calling for more radical changes to our business model. But these experiments are, well, experimental.

Whether such crowd-funding models of science journalism will be sustainable remains unknown. For now, more traditional models of publishing continue to predominate, and yet few would argue that that state of affairs is sustainable.

"My sense is that we're doing less with less," says Vergano. "Pieces could be a lot better if everyone had more time to think and talk. But the economics don't allow it."

Henig laments that she made roughly the same annual income as a freelancer 30 years ago—about $50,000, give or take—as she makes today. "It does help to know what the industry-standard rates are to ensure that you get paid decently," she says. But knowledge isn't enough when the industry standards themselves remain stubbornly stagnant. In her NASW presidency, Henig has vowed to try to find organized ways for writers to join forces against low pay. "We need enough people who will say, 'I can't work for a place that treats writers the way you do.'"

Something will eventually give way. Until then, writers' best—really only—option is to push back and stand firm whenever we can. Rigorously assess freelance pay rates and how they affect your bottom line, like you would any other dataset. Then ask both yourself and your editors the tough questions about compensation. You wouldn't do any less for a story. Don't short-change your own value as a professional writer, either.

7 Nice Niche: How to Build and Keep Up with a Beat

Knvul Sheikh

SCIENCE WRITING IS a great place to develop a beat, build expertise in an area of research, and report in depth on new, interesting work. Many beats are naturally circumscribed by scientific disciplines, such as anthropology, astronomy, energy, environment, or health. Other beats may be defined geographically (for example, the Lower Midwest or the Pacific Rim), by narrower areas of specialization within a scientific discipline (climate change, exercise science, or cancer biology), or both (environmental issues facing the American West). Still other beats center on the culture of science itself—for example, a journalist might focus on issues such as scientific publishing and reproducibility, science policy, sexual harassment in science, or science careers.

Specializing in this way is nothing new. The word "beat" has been used to describe this type of journalistic work for over a century. It began showing up in the late 1890s in the US to describe the different sections of news that reporters focused on, writes Michael Quinion, a former BBC studio manager and freelance reader for the *Oxford English Dictionary*. It's hard to establish the word's origins, but Quinion notes that it might have derived from the way policemen's feet hit the ground while "beating the streets" on their routes. Reporters similarly "beat a path" along familiar topics. They also gathered stories on foot at the time, which may have encouraged use of the word "beat" in journalism.

Carving out a beat in science journalism, in particular, offers numerous advantages. Compared with reporters who consider themselves generalists, beat reporters develop a deep understanding of the context of new scientific findings within a specific area. This contextual knowledge sharpens their ability to judge when a new finding is newsworthy and when it's simply a small step in the field, says Tina Saey, a senior molecular biology writer at *Science News*. Over time, beat reporters also build up a well-curated list of experts in their field of focus—people they can call on to comment on a new study or provide context for breaking news. Those connections can also enable beat reporters to find out about new research before it's published and before other reporters get the scoop, Saey says.

Picking a niche can also be a smart career move. Honing a specialty helps freelance writers build their business, becoming go-to reporters who editors can turn to for credible stories in a specific area. "Beats are really beneficial for building up your clips and your reputation in a field, and that makes it easier to pitch editors stories, because you know exactly what's interesting that's going on in that world," says Christine Yu, a Brooklyn-based freelance sports-science reporter.

Going to the Sources

At its core, beat reporting is about covering stories in a certain area often enough that knowledge of the territory, its most interesting aspects, and its denizens becomes second nature to the reporter. Reporters trying to build a beat can fast-track this process by cultivating an extensive contact list—sources they can reach out to for background information for stories they're considering pursuing, as well as for expert comment on assigned stories.

Yu suggests writers start by seeking out researchers cited in any study they're covering. Reporters can then grow their list by asking sources they interview to recommend other people with related expertise. Reporters can also reach out to public information officers at universities and other research institutions for help in adding to their list of experts, she says.

It's also a good idea, while reporting any story, to lay the groundwork for future projects. During interviews, query sources about who else is doing interesting research in their field, beyond those whose work is most relevant to the story at hand. And, Saey suggests, "If you call someone for comment on someone else's paper, spend a little time talking with them about their own research and ask them to keep you in mind when they're publishing something new. That probably works only 10 percent of the time, but that's still a significant percent—and as you expand your network, you'll have more and more people who are keeping you in the loop."

Periodically checking in with sources via email to learn about new developments in their work or in their field is also key to strengthening connections and digging up new stories. Just don't overdo it—emailing once a year or so without a particular story is about right, says Dina Fine Maron, a wildlife crime reporter at *National Geographic*. "People have limited time and don't want to feel taken advantage of," she says. When doing these follow-ups, be specific about what you're looking for and remind them of a connection you have, such as a story you previously interviewed them for or a study of theirs that came out afterward, Maron suggests. "Say something like, 'Hey, I see you did X. I'd love to learn more about that and also pick your brain about what else is exciting you at this moment.'"

Scientific conferences are also a good opportunity to cultivate sources on a beat. Journalists should take the opportunity to meet the presenters, who are typically important people in the field, and talk to graduate students and postdocs, who could be future leaders and sources in the beat. Journalists can approach a scientist if they find a presentation particularly interesting and let the researcher know they'd like to chat in more detail about the work at a later date. Attending poster sessions at conferences can also give reporters an idea of which preliminary research is gathering attention, who likes to chat with journalists, and which researchers are able to explain their work in a way that's useful for journalistic purposes, Saey says.

"For me, scientific meetings are basically like a one-stop shop," she explains. "You have sources, you have material, and you have potential story ideas all in one place."

Another way to find and follow potential sources on a beat is through social media. Following sources on a social media platform can help beat reporters find out what news or papers researchers are sharing, Maron says. Social media management tools, lists, and advanced search features are all useful for customizing who and what to follow online.

Staying on Top of Trends

As they become immersed in their areas of focus, good beat reporters go beyond what individual sources are doing in their fields to connect new research to the bigger picture and uncover stories about broader trends. The classic way to stay on top of new research in a beat is to sign up to receive press releases and table-of-contents lists for both high-profile and lesser-known journals in the field. But keeping up with these subscriptions can quickly become overwhelming, so beat reporters come up with ways of restricting the time this takes every day. Some writers recommend reading alerts in weekly batches. Others take an hour to check notifications and new developments every morning. "Basically, I've learnt the value of skimming," says Jessica Fu, a food and agriculture reporter at *The New Food Economy*.

Reporters can also deepen their immersion in their beats by scouring trade publications, Fu says. For example, she got an idea for a story on invasive black carp causing problems in the Mississippi River Basin after reading about incentives to catch and sell the fish in the publications *Seafood News* and *Undercurrent News*.

For beats that rely heavily on new studies as story pegs, like chemistry or medicine, reporters can look at embargoed news and get a pretty good idea of what their week will entail. Writers on these beats can also get scoops by combing through preprint archives for new research findings.

Journalists covering the environment or wildlife beats are more likely to focus on policy debates. Instead of checking journals for new studies, reporters on these beats may track advocacy groups or nonprofit think tanks such as the Humane Society or the Environmental Investigation Agency to find out what's important in their field, Maron says. Setting up Google alerts can also help reporters monitor when new information is shared by relevant institutions or when specific policies are mentioned on the web, she adds.

Akshat Rathi, a senior energy reporter at *Quartz*, recommends notetaking apps like Evernote or Notion.so for jotting down news alerts, press releases, and other findings that catch your eye when you're doing background reading. Rathi also periodically sends out a newsletter to keep his sources updated on new features he's reported, and often includes surveys to poll them for other trends they'd like to see him cover.

Fitness and sports-science reporters like Yu often rely on a different method for finding trends that readers may be interested in: Instagram. "There's a pretty big and active fitness community on Instagram, so if

I see something cropping up more and more frequently, that's something I'll take note of." To follow up on trends she's spotted, Yu then turns to Facebook and other social media platforms for further research or to find sources who'd be willing to share their experiences with her.

Keeping It Fresh

Beat reporting's advantage—and also its challenge—is the necessity of continuously evolving and finding new story angles on the same topics, says Nicole Mortillaro, an astronomy reporter at the Canadian Broadcasting Corporation. Mortillaro found that as announcements of the discovery of Earth-size planets orbiting other stars became more regular in recent years, she experienced a phenomenon that she calls "exoplanet fatigue." But finding newer angles, such as how scientists were exploring exoplanets by studying their atmospheres, helped her overcome feelings of apathy from reporting on similar topics over and over again.

When journalists switch beats, however, they run into a different problem: They often have to retrain themselves to find unique story angles in their new field. In Maron's case, when she switched from writing about medicine to covering wildlife poaching, she relied on her old skills to critically analyze new animal-trafficking policies, rebuild her network of sources, submit FOIA requests, and piece together enterprise stories.

Reading other writers' work in the same beat also helps reporters learn to ask better questions and think about their beat in new ways, Fu says. "I think that one of the best ways to keep myself sharp is to read other really good work," she explains, "because that encourages me to write about things I hadn't written about before and inspires me to be curious about new subjects."

8 A Conversation with Amy Maxmen on "How the Fight against Ebola Tested a Culture's Traditions"

Amanda Mascarelli

WHEN AN EBOLA epidemic swept through West Africa in 2014–2016, resulting in the deaths of more than 11,000 people, I was gripped by the absorbing stories that were being reported from the ground by journalists like Amy Maxmen, who wrote more than 15 pieces on Ebola from Sierra Leone—ground zero for the epidemic—for outlets like *National Geographic*, the *Economist*, and *Newsweek*.

We journalists often long to put ourselves in such intense reporting situations, because this type of shoe-leather journalism brings us closer to understanding our world, our sources, and our humanity. In a time of dwindling budgets, with fewer and fewer publications subsidizing reporting trips, Maxmen's opportunity (through a grant from the Pulitzer Center on Crisis Reporting) to travel to the region during the height of the Ebola epidemic was rare and enviable. The desire to report from the field on important, human-driven stories is the reason many of us became journalists.

Maxmen took three trips to Sierra Leone, extending her first stay from 10 days to a month over the holidays because she was so moved by the number of important stories that she felt needed to be told. She recalls emailing her editors in the United States on Christmas Eve with urgency about news that she thought needed to be covered and feeling annoyed when they didn't immediately reply. "Where I was, nobody was talking about Christmas," she told me.

Maxmen's story "How the Fight against Ebola Tested a Culture's Traditions," published on January 30, 2015, as part of a series of Ebola pieces she wrote for *National Geographic*, puts a human face on the cultural crisis that faced many in West Africa as the outbreak raged on. Maxmen is a masterful narrator, using vivid storytelling in subtle ways to put us right into the unimaginably sad circumstances she encountered. Here, she shares the backstory behind her reporting in West Africa and the lessons that have shaped how she approaches her work.

AMANDA MASCARELLI: What was the genesis of your reporting on Ebola, and how did the Pulitzer Center travel fellowship come about?

AMY MAXMEN: As the outbreak was escalating in 2014, I became increasingly obsessed with covering it on the ground. I'm very interested in infectious disease, and I had been to Sierra Leone and cared deeply about the country.

When I apply for Pulitzer fellowships, I like to assure them that at least two outlets are interested in stories that stem from the travel. But that was really hard in this case. Most of the outlets I usually write for were turning my pitches down. My speculation is that it had a lot to do with liability—would they be on the hook for me if I was suspected of having Ebola, which would be really expensive. So firming up the assignments turned out to be tough, but I was desperate to go.

I jumped into the reporting by writing a piece for *Newsweek* on how poverty, slavery, and conflict fueled the Ebola outbreak. That reporting helped me come up with other story ideas. One, about treatment made from the blood of Ebola survivors, was commissioned at the *Economist* in early November, and finally, on November 20, an editor at *National Geographic* said he was interested in a piece about how traditions around death changed as a result of Ebola. The Pulitzer Center agreed to fund my travel, and I booked my ticket. Two weeks later I was there. I had expected to be gone for like 10 days. But it was so interesting. It just felt so important. And there was so much news everywhere. I extended my plane ticket twice and I didn't come home until January 1, so I was there for the month of December 2014, and returned for February and June 2015.

How did the idea of covering burial rituals evolve between you and your editors?

While I was still in New York, I talked to sources from Western NGOs and agencies like the CDC who told me about how Ebola was spreading through unsafe funeral rituals. They said that Sierra Leoneans continued to bury their loved ones through traditional practices that involved a lot of touching. At least one source said that Ebola spread at one funeral because people drank water out of a corpse's mouth. This story had attracted some media attention. So, they were hoping to encourage people to change their practices by making so-called safe burials more welcoming.

When I arrived in Sierra Leone, however, I found that this wasn't the whole story.

Everyone I talked with in Sierra Leone told me they had never drunk water out of the mouth of a corpse. So it's not impossible that this really happened, since there are a lot of traditions in the country that belong to various groups, but it was clear this story did not represent a norm. Instead, people mainly told me about how their loved ones had been taken away from them within the first six months of the outbreak. And that generated distrust that was hard to undo as the outbreak progressed. People would take their sick child to the hospital and then never hear from them again, have no idea where they were—they're just gone.

So as word of these experiences got around, people resisted burial workers because they worried about their loved ones being tossed into unmarked graves. And people were reluctant to send their family members to the hospital because they worried about not ever hearing from them again. If I had just reported this from the US, I would have missed this very important piece of the story.

What is your method for launching into your reporting when you're planning a trip to someplace like Sierra Leone? How do you figure out who to interview and how to get around? And how much of your reporting plan is set up in advance?

Typically on these trips, I like to have enough in place so that I know what I'm doing for the first 24 hours. I know how I'm getting from the airport to the hotel, and I know who I'm visiting for an interview. I call a few translators before I leave so that I have a sense of how well they speak English and if we're likely to get along. I also go with lists of phone numbers for sources that I've gotten by digging around online or in calls with experts. I'll also have phone numbers for friends of friends. And I say that loosely. I met the paramount chief of eastern Sierra Leone through an Uber driver in Brooklyn. I'll also have thought through sticky situations so that I can try to prevent them and have ideas on how to respond. Then, once there, I put all this in the background and let intuition guide me.

But I go knowing that the story and the people I speak with will change as soon as I hit the ground. I find that the best people who can tell you about the country you are visiting are the people who live there—especially when you're talking about how the culture responds to an outbreak. So I'll meet with my source at CDC who I have lined up ahead of time, and I'll listen to their take on the local culture. But a parking lot attendant outside of their hotel might be a more important source on this topic.

Sierra Leone is a really safe country, and most people speak English in the capital, so I would just walk around the neighborhood where I was staying and strike up conversations. People are curious about me, and I'm curious about them.

What safeguards did you put in place for your health and safety while you were reporting on Ebola?

I did my homework beforehand. I talked with infectious disease experts, and learned how Ebola spreads in the late stage of the illness and just after death, when viral levels are high and coming out of bodily fluids, in sweat, blood, or diarrhea. That's why health workers are at high risk. So I made the decision to not go into the "red zone" of Ebola treatment units. To do that I'd be wearing a Tyvek suit, and couldn't have an interview with patients anyway. There was no point. Photographers were getting great shots, so I could imagine what it looked like. It wasn't just my life I was worried about, but also if I got Ebola I'd be jeopardizing the time and safety of everyone who would have to deal with me after that point.

In the early stage of Ebola, people don't seem to be very infectious, but you still don't want to risk it, so you don't touch anyone. That's not hard if you're an outsider staying in your own room in a hotel, as I was. Another step I took was to hire a driver rather than take taxis, because a taxi might have been carting a person to the hospital just before you got in, or someone could hop in to share the ride. So when I met a person recommended to me by another journalist as a driver, and decided I liked and trusted him, we made a deal that I would pay a good fee and in return I would be the only person he drove around.

Another key was to not get sick, just generally, because if I got a fever then I'd have to worry that I might have Ebola. So I got a flu shot. I took antimalaria pills. I used hand sanitizer constantly. I was extra cautious about avoiding any food that might be contaminated. And I didn't get sick. Beyond that, I felt free to walk around and talk with people.

When you're reporting in the field, do you go home and translate your notes and jot out scenes? Or do you wait until you're drafting a story? Tell me a little bit about your process.

If I'm being good, I go back to my room and write down the most memorable things. I have a file on everywhere I've been, titled something like "Diary from Sierra Leone." And I just let myself be free-form with it. This helps me remember the most emotionally gripping situations. I also take a ton of photos with my iPhone and I'm not precious about it—I'm just snapping a shot so that I can jog my memory later. So I'm obsessive about downloading those, and then copying them onto a couple of USB sticks, along with the text file. This way, even if I don't get to write a story for two months, I can go back and look at those pictures and my free-form diary and put myself back in that place.

I thought it was powerful the way the image of the health workers in the Tyvek suits kept recurring. Who were they? Were they a mixture of locals, or were they outsiders? Was there a strategic reason why you left that perspective out?

I don't even know if it was intentional. But it works, because I wanted to convey how they were seen by people. Here's a bizarre stranger, the opposite of who you'd want to see when you're in mourning. They're national, as in from the country. But sometimes they weren't part of that community. They were usually hired by hospitals or NGOs. And I did talk to a lot of them. I guess there are no direct quotes because in the end you have to choose the most powerful quotes, and those weren't them, at least for this piece. But I got information from them—like, for example, there's an anecdote about how they would dress a corpse in a dress or they would put a necklace around them before they put them in the bag, if the family desired it. I did quote burial workers in another piece I wrote for *Newsweek*, and that directly addressed the main thing they complained to me about—a lack of payment.

You did a good job of avoiding characterizing the people there and their rituals as strange or different from "us." Were there any cultural sensitivities that you had to be particularly attuned to while reporting and writing this?

Thanks! I feel like sensitivity comes from having in-depth conversations with people. It's the only way I'd cover a story like this; you have to be there. Which is unfortunate because nobody has travel budgets anymore. But if I had just called Western NGOs and agencies like the CDC or the UN, I would not have gotten the story. When you talk to people, and you start hearing their stories, they don't seem so weird anymore. They usually make the same decisions you would in their circumstances. So you really need to be there, and to be fearless about talking to people. I find generally people like to talk about their lives. If I'm on a call with a scientist to talk about science, I keep it to 30 minutes. But if I'm asking somebody about something traumatic, I don't limit that time. I might go to their house, or somewhere else where they're comfortable.

Did you learn any major lessons while reporting this, or did it change the way you approach your reporting or writing?

I learned a lesson by making a terrible mistake. I talked with a lot of people who had survived Ebola, and maybe as a result I started moving through these interviews a bit fast. Reporter mode kicked in and I was looking for the details I needed to convey their experience accurately. So for instance, if someone said that they'd step over dead bodies to go to the bathroom in the Ebola ward, I would follow up by asking how many bodies they stepped over. And at the end of the interview I would be like, "Thank you so much, see you later."

I didn't realize that was wrong until I met up with one of the survivors again on other visits to Sierra Leone in February and June of 2015. One afternoon, this survivor told me that after he relived the experience while talking with me, he felt terrible for days and didn't leave the house.

I had no idea that I was causing this pain, and the only way to fix the situation now is to not repeat that mistake, and to share the story so that other journalists don't do the same.

What can be confusing is that the adults I interview want to share their experience because they want the world to know what they are going through. And it's my duty as a journalist to collect the facts needed to be sure their stories are accurate. But now I have a new strategy that I hope lessens damage caused by conjuring up the past. If you interview someone about a traumatic experience, spend a good deal of time at the end of the interview talking about positive things in their life, the things that give them strength. For example, I just returned from Jordan, where I met with a number of Syrian refugees. One woman was in tears as she told me about how the Syrian regime had burned her son alive and had arrested and beaten another son. I tried to pivot by asking how she's so resilient. She mentioned her husband, and I asked how they met, and then about the farm they lived on, and the cheese she made. We talked like this, and when I left we were kind of laughing. I need to give credit to my translator who really helped lighten the mood. There is nothing I can do to heal wounds like this, but at the very least I can try to not be a jerk.

What Makes a Science Story and How Do You Find One?

PART 2

What Makes a Science Story and How Do You Find One?

YOU'VE FOUND SOMETHING you'd like to write about. Perhaps a paper in a scientific journal or a talk at a research conference has caught your attention, or you've read a local news story and a question popped into your mind that surely has a scientific answer, or maybe your neighbor has told you something intriguing. However you've found the idea, if it makes you mad or curious or frustrated or awed, or brought out any other strong emotion, that's a good clue that there's a story lurking.

Odds are, though, that what you've got so far is a *topic* and not yet a *story*. To refine it into a story idea, you need to be able to answer questions like these: What's timely about it, the peg that makes it something people need right now? What will your angle be? That is, what lens will you apply to the topic to organize the information and make it meaningful to readers? Are you sure that any scientific claims underlying the piece are solid? Will your piece be a news story, a long feature, an infographic, a profile, an essay, an op-ed, a Q&A? What's the thread readers will follow through the piece? Will it be a narrative following key characters, the evolution of an idea, a controversy with unfolding arguments? How will you structure the piece—where will it begin, what ideas will form the middle, and how will it end?

Answering those questions will require an initial round of reporting. Then, when you have your mind wrapped around what the story will be, you need to explain it to an editor, infecting them with your excitement and convincing them that you can deliver the goods—that you can find the right sources, synthesize what you learn, and pull the story off in a way that fits their publication. In other words, you need to write a barnstormer of a pitch.

Pitching is one of the hardest tasks in journalism. While it's absolute lifeblood for freelancers (who depend on their pitches to get assignments and thus, eventually, paychecks), staff reporters also have to sell their editors on story ideas. And pitching is a particularly tricky skill to learn because pitches are hidden in private communications between journalists and editors, so a new journalist can't absorb the conventions by casual observation. In the chapters that follow, seasoned journalists analyze what makes successful pitches great. Further resources at *The Open Notebook* pull back the curtain by providing an online database of hundreds of successful pitches to dozens of different publications, which you can analyze for inspiration and guidance.

Finally, never forget the key role of persistence in pitching. A rejected pitch isn't a sign of failure; it's a sign of being a journalist.

Siri Carpenter

9 Is This a Story? How to Evaluate Your Ideas Before You Pitch

Mallory Pickett

WHEN A JOURNALIST is on the prowl for a new story, every yarn spun by a friend, every press release, every vacation adventure, and every quirky local news item can seem like the beginning of a great story idea. But most glimmers of inspiration turn out to be just that—transient inklings. Only a few will be real gems. The most successful freelancers can quickly sift through their ideas and see whether an idea deserves to exist as a story, and what kind of story to pitch. This evaluation is usually an intuitive process, honed over years of practice. But learning the right questions to ask about your ideas can help accelerate the learning curve.

The process always starts with a spark: Something seems interesting, and you want to learn more. "Interesting" can encompass a lot of characteristics. Francie Diep, who covers science for *Pacific Standard*, says when she scans her bookmarked sites she's looking for anything that's new, controversial, related to the news of the moment, or that brings evidence to a debate that needs it—and stories about cute or weird animals are always good, too.

Meera Subramanian, an award-winning freelance journalist and author, says she finds most of her ideas while reporting on other stories—by paying attention to moments when her sources bring up surprising ideas or her curiosity is piqued in some new way.

But no one's "interesting" filter is completely reliable. "Sometimes you think you discovered something new and it seems fascinating, and then once you look into it you realize this has been completely reported, everybody else knows about it, this was just a new fact for *me*!" Subramanian says.

Jennifer Kahn, a contributing writer for the *New York Times Magazine* and an instructor at the University of California, Berkeley School of Journalism, says students often struggle with this problem—trying to

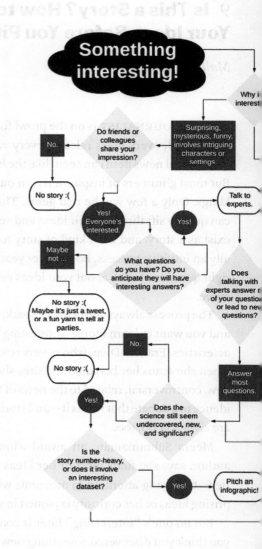

get past their own initial interest to evaluating whether an idea will be compelling to others.

"It's not enough for something to be surprising or interesting. You have to think about *why* it's interesting," Kahn says. "The number one thing, at least with students moving from intro to more advanced classes, is that people have stumbled on something they randomly find

Decision-Making Flowchart

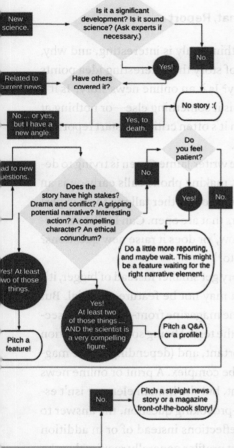

New science.

Is it a significant development? Is it sound science? (Ask experts if necessary.)

No.

Yes!

Related to current news.

Have others covered it?

No story :(

No ... or yes, but I have a new angle.

Yes, to death.

Do you feel patient?

No.

...ad to new ...uestions.

Does the story have high stakes? Drama and conflict? A gripping potential narrative? Interesting action? A compelling character? An ethical conundrum?

Yes!

No.

Do a little more reporting, and maybe wait. This might be a feature waiting for the right narrative element.

...es! At least ...wo of those things.

Yes! At least two of those things ... AND the scientist is a very compelling figure.

Pitch a Q&A or a profile!

Pitch a feature!

Pitch a straight news story or a magazine front-of-the-book story!

No.

Download a print-friendly version of this flowchart at bit.ly/TONflowchart.

fascinating—'Oh my God, it's a woman-run mosque,' or 'Oh my God, the falafel guy is actually from Syria'—and they are kind of curious about this. But there's no interrogation of what actually makes it so interesting to them."

Beginning writers are usually able to easily identify topics or people they want to learn more about, but they can sometimes get stuck on the surface—they don't know how to dig down to find out if there are fundamental questions about them that a lot of people would like to know the answer to.

To Find the Right Scope and Format, Report

Once you've determined that something truly is interesting, and why, the next step is deciding what kind of story this interesting idea points toward. Is it a magazine feature story? Is it an online news story? Is it a profile? A Q&A? An infographic? Or is it something else—or nothing at all? This moment of decision is when it's often critical to start reporting and get input from potential sources.

When essayist, editor, and science writer Jaime Green is trying to decide the scope and format of a story, making phone calls early on is an important step. She says it comes down to whether talking to an expert "opens more doors or closes the doors that are open. Can I talk to someone and say, 'Oh, OK, cool, I get it now,' or does it raise new questions? Are there more people I want to talk to?"

If this pre-reporting makes the story get smaller instead of bigger, it's probably an indication that the idea may not be feature material. But that doesn't mean it's not a story. Some magazine front-of-the-book sections might require that an idea pass the test of being some combination of interesting, surprising, new, important, and (depending on the magazine) quirky, but it doesn't need to be complex. A print or online news story needs to be new and significant, but the human element isn't essential. An essay requires a thought-provoking question, the answer to which will come from the writer's reflections instead of or in addition to their reporting. Successful scientist profiles generally require that the subject's personal story be as interesting as their research.

A Decision-Making Flowchart

All of this preparation and decision-making can be exciting—you never know when you're hot on the trail of your next big story! But it's also taxing and, for freelancers, unpaid. Developing criteria to evaluate ideas can make the process a little more efficient and economical. Every writer will eventually have his or her own ideas about the best way to do this, but learning from others' experience can help you get started. In this chapter, we present a flowchart created from interviews with the experienced writers mentioned above, as well as other journalists *The Open Notebook* has previously interviewed. The chart guides you through some of the questions you can ask yourself the next time your "interesting" alert goes off.

10 Sharpening Ideas: From Topic to Story

Dan Ferber

GEORGE JOHNSON WANTED to write about new developments in cancer research for the *New York Times*. But he needed to find a story that would let him do it. So in 2011 Johnson, a regular contributor to the *Times'* science section who's writing a book about cancer, cut a deal with his editor. He'd go to the annual meeting of the American Association for Cancer Research in Orlando, Florida, to see what he could learn from the world's top cancer researchers. And if he found a good story for the *Times*, they'd split his expenses.

As science writers, we learn about fascinating topics daily, and explaining that complexity is one of the joys of our work. But to sell the story to our editors, we need a good angle and often a compelling narrative approach. This can challenge even experienced writers, as I know from once-promising reporting that sits untapped (and uncompensated) in my computer.

Often, the key to finding novel, surprising, or controversial angles that nab assignments is to look at the story from different points of view, says Emily Laber-Warren, who has edited features for *Popular Science*, *Women's Health*, *Scientific American Mind*, and other magazines, and who directs the Health and Science Reporting program at the City University of New York's Craig Newmark Graduate School of Journalism. "The first thing is, don't give up," Laber-Warren says. "You have to be like a two-year-old holding a peg trying to make it fit into a hole. You have to turn it around."

Anatomy of a Story's Development

Johnson was still turning his ideas around when he arrived at the cancer conference, but he had a plan. He knew that two top cancer biologists had published an influential paper a decade earlier that laid out six hallmarks of cancer cells. He also knew that the two had just published an update of their classic paper, complicating the picture without disrupting its central principles. Johnson, who has also covered cosmology, saw that the two fields had developed similarly: Cosmologists still hew to the big bang theory, even as they've been forced to grapple with complications like dark energy and dark matter that paint a richer and more nuanced picture of the early universe. "I had the idea that the science of cancer was undergoing a big bang shift," Johnson says.

He envisioned a trend story for the *Times* that explored that theme. But he needed a news peg that offered a good excuse to explore the topic. An annual cancer research conference was not a strong enough news peg by itself, at least not for the *Times*, Johnson says. The recent update of the classic 2000 paper was a noteworthy development, but also not quite enough. "You want something unique that jumps out at that particular meeting," Johnson says.

A couple of days into the conference, Johnson attended a lecture by Harvard University cancer researcher Pier Paolo Pandolfi, who described the emerging role in malignancy of noncoding DNA, including pseudogenes and regions that encode microRNAs. Johnson realized that cancer biologists' new attention to noncoding DNA, sometimes referred to as "the dark side of the genome," was an apt example of a "big bang shift" occurring in cancer research. Johnson's trend story was shaping up, and he soon got a green light from his editor.

Next, Johnson had another problem to solve. "This was a really hard piece to structure," he recalls. He couldn't simply build his story around a narrative of the conference, as that wouldn't hold enough interest for readers (or for his editors). And focusing on a single character would have defeated his purpose of surveying the larger terrain of cancer research. Ultimately, Johnson pulled back his camera and described the decade-old view of cancer, then used the conference as a framework to introduce new complications, focusing on noncoding DNA's role in cancer development.

At the conference, he had attended talks that illustrated ideas he wanted to cover, and he later watched video recordings of important sessions that he'd missed. That legwork let him pull from Pandolfi's talk and several others, describing how healthy cells are conscripted to support tumors and how microbes in and on the body alter susceptibility to certain cancers. He included a few quick scenes from the conference to ground the story, and he finished with Harold Varmus, the director of the National Cancer Institute, speaking there on cancer's unsolved mysteries. "That gave it a satisfying structure," Johnson says. And the *Times* split the cost of his expenses for the resulting story, "Cancer's Secrets Come into Sharper Focus."

Chasing Your Tale

One way to sharpen a story idea—to nudge it from a *topic* to a true *story*—is to think through the ultimate structure the piece might take. Fortunately, settling on a good narrative approach helps make the pieces fall into place. There are lots of questions to answer, Laber-Warren says: Where will you go? Are you going to make it a travel journal (such as George Black's "The Gathering Storm," published in *OnEarth* in 2008, or David Samuels's 2008 *New Yorker* piece "Atomic John," a story that intertwines two narratives, one of them a travel journal)? Or will you do a narrative and find a person whose life you can tell it through (as in Darcy Frey's 1995 *New York Times Magazine* piece "Does Anyone Here Think This Baby Can Live?")? Or will it be a first-person experiential story (like Jenny Everett's 2004 *Popular Science* piece "My Little Brother on Drugs," or James Vlahos's 2005 piece, "Car Crashes . . . Criminals . . . Cancer . . . Black Swans? AAAAAIIIEEEH!," published in the same magazine)?

Or maybe your target magazine breaks broad topics into feature packages with separate, short pieces that have a common theme. For an issue story, if you can't find a central character whose tale is compelling enough to build the piece around, you can still find characters that individually represent an aspect of an issue, treating each aspect in a separate section, Laber-Warren says. For example, one section might focus on pros and another on cons, or sequential sections might cover ethics, science, and human impact. Such stories, which Laber-Warren calls "narrative-issue hybrids," are "much easier to find [than full-blown narratives] and . . . they're very effective." (Laber-Warren points to two classic examples—Rebecca Skloot's 2003 *Popular Science* piece "Sally Has 2 Mommies + 1 Daddy" and Bijal Trivedi's 2005 *Wired* piece "The Rembrandt Code.")

Sometimes characters can be found that represent sides of a controversy, and their respective narratives can be braided. Investigative medical journalists Shannon Brownlee and Jeanne Lenzer used that approach in 2011 for a piece they wrote for the *New York Times Magazine* on a raging debate over whether the PSA blood test for early prostate cancer helped or harmed patients. Through their reporting, the two had learned that the PSA test, lauded by many doctors for saving lives, actually spots many prostate cancers that are too slow-growing to cause harm. A positive PSA test then spurs doctors to treat patients aggressively, including surgery to remove the prostate gland. This in turn causes lasting impotence and incontinence in thousands of men whose slow-growing cancers would never have harmed them. Their piece, "Can Cancer Ever Be Ignored?," was driven by one central question, Brownlee says: "How the hell did this happen?"

To tell the tale, Brownlee and Lenzer wove together two narratives: one of a doctor who developed and popularized PSA screening, and another of a different doctor who says the test leads to overly aggressive treatment of benign, slow-growing cancers. Separate sections explore these two aspects of the story, and a third tells the story of a PSA-positive patient who agonized over whether to treat his cancer aggressively.

Narrative Casting Call

Finding a compelling narrative for an issue story can take a lot of careful reporting. A few years ago, Barry Yeoman, a Durham, North Carolina–based magazine journalist, learned that a 1993 Supreme Court decision had shifted the power to judge the quality of courtroom science from juries to judges, most of whom had little or no scientific training. This in turn gave corporate defendants in product-liability lawsuits powerful ammunition to attack the qualifications of plaintiffs' expert witnesses and disqualify them from even testifying. After reporting widely and deeply, Yeoman concluded that the Supreme Court ruling "was devastating to consumers with legitimate cases," and realized that his story would have to drive home that conclusion.

Then came the question of how to find the right narrative. Yeoman wanted to highlight a lawsuit that illustrated the problem he'd uncovered, so he interviewed legal experts and collected cases. He applied a number of filters to find just the right lawsuit. Because the story was about a dry topic, Yeoman says, "You had to care about the victim." Whatever case he chose had to involve a scientist whose work was unassailable, a science that was approaching consensus, and a corporation using its legal firepower to disqualify the scientist from testifying.

Yeoman ended up building his story around a case of a 13-year-old boy who had taken a widely used antidepressant, started behaving erratically, then hanged himself. When the boy's parents sued the drug company that sells the antidepressant, the plaintiffs' lawyers called an eminent scientist who had studied the relationship between SSRI antidepressants and suicide in children. The scientist's credentials were strong. "He was not a wild-eyed radical. He prescribed SSRIs himself," Yeoman says. But after the drug company attacked the scientist's work as "junk science," the judge barred him from the trial and the case was dismissed.

Yeoman knew he had the outline of a powerful narrative that drove home his conclusions. Court documents from the parents' lawsuit, including pivotal court transcripts and depositions from the boy's relatives and a teacher, provided a wealth of narrative detail. Yeoman was convinced the case presented a deeply compelling story. About two weeks into his reporting, he floated his narrative approach to his editor, who signed off on it. Yeoman's story, "Putting Science in the Dock," was ultimately published in the *Nation* (not the magazine that originally assigned it). It braided the tale of the dead boy and his family with the tale of the expert witness and his science, bringing to life how the Supreme Court decision had led to misuse of science in the courtroom, tilting the playing field in favor of corporate defendants.

Place as Character

For some issue stories, it's worth asking if the story of a particular place, rather than the story of a central character or two, best illustrates the issue. The place essentially becomes a character in the story, Yeoman says. For example, in a piece he wrote for *Audubon* in 2009, investigating whether green jobs were a reality, Yeoman visited Newton, Iowa, where Maytag factories that once employed 4,000 people to build washing machines had shut down, but two new wind-turbine-manufacturing plants had sprung up.

Focusing on such a place can help you choose who to feature in the story, Yeoman says. In the green-jobs piece, "Work Plan," he featured a former Maytag assembly-line worker who now worked for a windmill-blade manufacturer that had set up shop in Newton. Yeoman also included the voices of several other Newton residents, including an economic-development official who had helped recruit the new employer, the plant's manager, and the town's mayor.

Thinking Like an Editor

Since finding a good narrative for a story can take a fair amount of reporting, and time is money for freelancers, the initial reporting that's needed to pitch a story can be a financial gamble. To what extent does a freelancer need to detail the story's narrative to nail down a feature assignment?

It depends. "I'm happy to get the topic assignment and turn it into a narrative," Yeoman says, adding that his editors trust him enough to assign the story before he has found the narrative. But he concedes that this is a luxury of experience, and less experienced feature writers may have to do more work up front to nail down an assignment. "As trust and experience go up, the need for a specific narrative in the proposal process goes down," he says.

Laber-Warren recommends that writers suggest a narrative approach in their pitch. "It's important to have an idea you can throw out there," and be sure that it's the sort of piece your target magazine actually publishes. But she also recommends that writers signal a certain flexibility. "You could say, 'I could do this as a profile of this central person, or I could do this as a narrative in this or that way,'" she explains. "That's a great way to go. It shows you're flexible and shows you're thinking like an editor."

11 Critically Evaluating Claims

Megha Satyanarayana

OUR INBOXES ARE full of them—press releases, pitches, and other media calling some scientific event "a breakthrough," "a game-changer," or "a paradigm-shifter." Scientists, investors, and analysts flood our social media feeds, cheerleading a preprint or singing some company's praises, even when there is little to no data to back up their claims.

Figuring out whether something is newsworthy can be hard. But, as science journalists, we need to examine such statements and decide: Is this worth covering? If so, how do we do so objectively, without accidentally becoming a mouthpiece for hyperbolic claims?

What's at stake is significant. Information comes at us like a fire hose on full blast, and social media algorithms have made it easy for lies to spread faster than truth. For example, antithetical claims have continued to try and sow doubt around the causes of climate change. And misinformation problems have only worsened during the pandemic: In a 2021 Kaiser Family Foundation poll on false statements about COVID-19 vaccines, researchers found that 78 percent of people either believed or weren't sure about at least one of the claims. For journalists on tight deadlines, sifting fact from fiction can sometimes feel impossible.

But coverage lends credibility, which matters immensely to readers. Some of the things we write can profoundly affect people's actions, especially in health and medicine, says Rosie Mestel, the executive editor of *Knowable Magazine*. "There are a lot of people who are desperate and very sick," she says, "and you have to be very, very careful that you're not going to be misleading people and overplaying things."

To cut through the murkiness and hype, science journalists need to vet the information and sources they come across and be on the lookout for red flags. Also essential is understanding our own biases—what we wish to be true, and how that plays into our decision making. Here, both skepticism and self-awareness can be key.

Journalists have the power to tell or not tell a story—and how we dissect claims plays into that power, says Ashley Smart, a physics journalist who is the associate director of the Knight Science Journalism Program at MIT and a senior editor at *Undark*. "We owe it to our readers and to the general public, and even to our sources, to be thoughtful in what we decide to cover, and to make sure that it's worthy of the platform that we're giving it," Smart says.

A Critical Eye Pays Off

Information comes to us in many ways: press releases, videos, reader tips, research papers, conference presentations and posters, and more. But not all of them will package data and information in the same way.

Press releases are as much about getting attention for the institution, the company, or the researcher as they are about the research, says Janet Stemwedel, a philosophy professor at San José State University who has written on the topic of evaluating claims in research. A lot of nonprofits and institutions use press releases and media they have developed in-house to raise funds, so the tone and language will almost always be optimistic and positive. They may even oversell the value of the research.

Jonathan Wosen, a biotech reporter at the *San Diego Union-Tribune*, laughed when we discussed a recent example: A headline on a press release that gushed about "breakthrough pre-clinical data."

That's usually an oxymoron, he says: "Preclinical" means the experiments were performed in animals or even cells. What works in animals doesn't often work in humans. As Mestel says, scientists have often cured mice of cancer or destroyed a plate of cancerous cells with some treatment, only to have the treatment fail in people.

Press releases are one form of communication that can be rife with bias, but they don't stand alone. Here are some general tips for evaluating information in your quest to report objectively.

Wosen relies on a few questions to help him parse the many, many health and medicine pitches and press releases in his inbox, but the questions are applicable to any information you are evaluating:

Where's the evidence? Some press releases will link to data slides or peer-reviewed publications, but not always. It's up to us to ask.

Is the study in animals? In people? Rarely, animal studies are groundbreaking. But more often, they aren't.

How big is the study? The more people, the better for statistical analysis.

What are they measuring, and is it a valid way to measure it? This is especially important in clinical trials, where, for example, a decrease in some biomarker the company is measuring has little effect on disease outcome.

Many of these questions can be answered by looking directly at the data the information is based on. Smart does this to dig out the quantitative truth behind the qualitative claims in those press releases—for example, the numbers behind phrases such as "vastly improved x," "increased y," "greater efficiency of z." It's not always a straightforward process: Journalists may have to do a little math to figure out exactly how much something changed, and to determine if that change is really significant and meaningful.

Asking for Help

Sometimes, though, we don't have a study or paper to look at. During the COVID-19 pandemic, for instance, drug companies and diagnostic test makers, not to mention politicians and nongovernmental organizations, have made statements about products and research without releasing data or providing support for their claims. The data from companies will sometimes come out later, in the form of a peer-reviewed paper or in an earnings call, but the news is breaking now, and journalists have to decide whether and how to cover it. Even in the absence of detailed data, Stemwedel says, reporters can ask study authors how they designed, conducted, and analyzed their experiments. Sometimes, a follow-up story might even be warranted, after the full data are released.

As Wosen notes, one of the best ways to confirm the validity of a claim or the newsworthiness of a piece of information is to hit up a source who wasn't involved in the work to help out. "Those experts can sometimes save you a lot of trouble and help you identify whether something's a story or not," he says.

While at *Physics Today*, Smart (who is a member of *The Open Note-*

book's board of directors) says he and his colleagues routinely reached out to scientists who were not part of the research to ask what amounts to two basic questions: Are people in the field calling this work important? Will whatever claim the scientists are making stand the test of time? "We basically do our own little peer review," Smart says. This type of vetting is standard practice at other science publications.

Finding these folks can be challenging for a reporter new to a beat, or a journalist covering something they don't have as much experience with. Smart says one place to start is the references at the end of the paper, if available. I've found that, paper or no, a quick call to my local university can also turn up people who can help. So can searching through older news stories on the same topic. I frequently turn to my colleagues as well. When I worked at *Chemical & Engineering News*, it was common for us reporters to ping one another for sources. Not all colleagues share, but many do.

Some of these early assessments will turn up disappointing answers: Maybe the story just isn't there. If that's the case, Mestel says, "it's OK to abandon ship."

Turn the Lens Inward

Journalists have to ask tough questions, dig for information, and even question our own belief systems as we interview, research, and write. Our own biases can influence our eagerness to cover something, whether it's a fascination with new technology or gadgets or a dire report on oil spills that plays on someone's love for animals. We need to do gut checks, contextualize the evidence provided, and add caveats and nuance to temper expectations.

For example, journalists love reporting on foods like chocolate, says Alice Lichtenstein, a nutrition professor at Tufts University who helps dispel myths about food. But nutrition reporting is often full of holes, and journalists often bring too little skepticism to their coverage of food research. It's hard to pin down why, but she thinks it has something to do with the idea that because we all eat, we all think we are experts in doing it.

Another common reporting mistake can occur when journalists don't fully understand the nature of the study they are reporting on. A lot of climate change studies, for instance, are based on modeling of different events and drawing conclusions based on those models. Not understanding how the models work, and what their shortcomings are, can lead to overselling or underselling the research. And in biomedical or clinical research, it's important to draw distinctions between interventional studies, where researchers change people's behaviors or treatments and look for effects, and observational studies, where they just look for patterns in what people are already doing. "With an intervention study, it's essentially cause and effect, where with an observational study, you just look at associations," Lichtenstein says.

Journalists will always benefit from understanding statistics better, and the importance of study-design issues like sample size. (Bigger studies are generally better.) Outside experts, too, can help if more complex statistical methods crop up; Wosen has developed a relationship with a source who at times weighs in on what he's considering reporting on. Lichtenstein says she has often served that expert/evaluator function, and sometimes she has told reporters the study isn't newsworthy.

And of course, it's important to navigate conflicts of interest: Ask sources who funds the research? This is especially relevant to climate change and environment reporting, where organizations with a stake in the outcome of climate change mitigation can be prone to hyperbole.

Sometimes, a conflict of interest (COI) becomes part of a story. I once wrote a story about weight-loss drugs in which one of the academic researchers I spoke to had worked previously for a company we were talking about. I chose to keep him in the story and highlight that relationship because he is truly one of the experts in the field, and I made sure to include his comments on the field at large, as well as other companies.

But sometimes, a COI means a source has to be excluded because they can't be objective. Stemwedel says excluding a questionable source is about maintaining credibility with your audience.

Yet, for all our due diligence, evaluating claims isn't foolproof. Gale Sinatra, a psychologist at the University of Southern California who specializes in STEM education, offers this tip: Remember that search engines are built on preferences. Every time you search something, it tells the algorithm you're interested. As you keep searching, you get more of the same. That's how misinformation campaigns get rooted—a search for a conspiracy theory brings up all kinds of links, which teaches the algorithm that this is what you want. If you want to evaluate a claim, she says, or the person making the claim, do your research in your browser's private or "incognito" mode.

Perhaps, for example, you want to double-check a source's potential conflicts in a story about online wellness apps. Since you've been using your browser to research apps, when you search for your source, the algorithm may give you more narrow results related to just wellness. But if you try an incognito search on just that person, you might get different results, including criticism of that source, ties to companies that make apps, or a news story involving that source that may call their credibility into question.

It can be discouraging to find yourself navigating baseless claims in the search for news. After all, Mestel says, many people get into journalism because they want to get to the bottom of things. It's not always easy, but in this climate of misinformation and bad actors trying to stir up trouble, digging a little deeper into superlatives and claims isn't just about credibility, but also personal integrity, and being honest with yourself.

But Mestel can always remind herself of what motivated her to become a science journalist in the first place. "I want to understand what's going on in the world," she says. "I want to write about what's true."

Kate Morgan

BACK IN THE spring [of 2018], an editor emailed me to ask if I'd take on an assignment about comedian Amy Schumer's recent movie, *I Feel Pretty*. I don't write about entertainment or Hollywood, and by that time the film had already sparked reviews and op-eds in dozens of publications. The editor and I worked to find a different angle: a scientific one.

I spoke to experts and read studies, and I also sat through all 110 minutes of the movie. The piece I ultimately wrote was part neuroscience and part humor, and my editor was thrilled. Not all science writing has to be deeply academic or focused on dense concepts. Science writing can be crowd-pleasing, and shareable, and even sarcastic or funny. And it can belong in publications whose focus might seem far removed from science, such as magazines centered on fashion, business, food, public policy, sports, parenting . . . or just about anything else.

For a freelancer, finding a scientific angle on a trending news topic can make a pitch pleasantly unexpected, and more likely to pique an editor's interest.

"[Science is] one more way of understanding the world and unpacking a current event on another level beyond the 'who, what, where, when, why,'" says Cari Romm Nazeer, a former editor at *The Cut* who now heads up *Medium*'s service and advice section. "When you take a story and try to extract science from it, it can sometimes make that thing a little more relatable, and it can be a fresher way of getting into a story that readers aren't used to."

You also, in my experience, don't have to be a scientist—or even consider yourself a science writer—to write about science. Instead of writing about a mental state they're in, a personal essayist might examine why their brain is doing what it's doing. A music journalist might talk to experts about the harmonics and acoustics of a new album, or how the science of sound impacts what does and doesn't make the charts.

Nazeer argues that there's at least a little bit of science in everything people "do or touch in our day-to-day lives," and becoming attuned to it could open countless avenues for writers to turn ideas into stories. If your goal is to break into science writing, the starting point is simple: Consider an interesting question or examine an intriguing character. Then, you're well on your way to crafting a piece that tells a great story, with science to back it up.

Identify the Science

Some stories about science don't start out that way. Sometimes, the science just shows up, says Andrew Zaleski, who's published features in *Popular Science*, the *MIT Technology Review*, and *Medium*. That was the case with his story about a shark fisherman, published in the summer 2018 issue of *Popular Science*.

"The theme of the issue I was pitching was 'danger,'" Zaleski says. "I'd heard about this fisherman, and I thought, 'OK, shark fishing is pretty dangerous. So, what can a shark fisherman tell us about science?' Is it smart, safe, good, bad to take sharks out of the ocean? What does this guy think?" Ultimately, the story became one about conservation and marine ecosystems.

Zaleski doesn't consider himself a science writer, but he knows that, often, focusing on the scientific aspects of a story can help him sell it to a particular publication.

"There are a couple of stories I'm looking at now that are science-y," he says. "But I also think they're just good stories. I'll dial the science up or down, depending on the publication that takes them. Sometimes the story is just there to convey the information about the science. Other times the science is just a companion to the story."

In October, Zaleski published a feature about the invasive spotted lanternfly in *Bloomberg Businessweek*. That publication may seem an odd choice for a story that quotes entomologists and horticulturalists, but it was a case, Zaleski says, of tailoring a science story to fit a target outlet. "The story ultimately has an economic angle," he says. "But in order to talk about why these bugs suck so much, you have to explain the science."

In that case, Zaleski approached a science story from a different angle, but it's also possible to do the opposite, with an idea that may not initially appear to have a scientific foundation.

Nazeer says there's a wealth of scientific questions surrounding the thoughts and events people experience in their daily lives, all just waiting to be asked—and written about. "Finding it just requires paying closer attention to patterns in everyday life that we don't necessarily think of as being science material," she says. "There are so many quirks of human behavior that can have a lot of psychology applied to them."

Nazeer points to one of her favorite stories, which she wrote in 2017 for *The Cut*: a piece that traded on the trope of why so many couples fight at Ikea. "That's a good example of taking something we don't think of as a scientific phenomenon and talking to psychologists about what's happening there." Nazeer spoke with experts whose research backed up her hypothesis that visiting Ikea together is bad for a relationship. "They all explained that trying to make decisions together and balance each other's needs and wants creates this kind of perfect storm and makes our brains want to get into an argument."

Many of the best science stories Nazeer has assigned, she says, come from writers looking for a deeper explanation of something seemingly insignificant. "I think the best thing you can do is pay close attention to idle conversations with your friends, or the weird things you find yourself thinking about," she says. "Those casual moments can sometimes be the spark for an unexpected idea. We had a piece a while back about the difference between cute and 'ugly cute.' It came out of the writer freaking out about a pug. You can take those things and treat them as phenomena worthy of inquiry, and get a fresh, fun, surprising story as a result."

Find Your Expert

Once you've identified the scientific inquiry hiding in a story idea, the next step is to find someone with actual credentials to explain it to you.

"It can be a doozy sometimes," Zaleski says. "I rely a lot on the sources I talk to. I wrote a story for *Popular Science* about a guy who's pursuing immortality, and I was finding people who work at aging-research centers, and I read this book that was like a crash course in biology. I probably spoke with 12 to 14 people before everything started coming together."

While this is, admittedly, the time when having a science degree would come in handy, it's possible to distill complicated concepts by just being a good reporter. Scientists often respond to an honest, empathetic interviewer just like any other source, and asking detailed, informed questions about their research may encourage them to open up. It's also helpful to know what you don't know, and not be afraid to ask. Most importantly, you have to be willing to do the research. In order to explain the science to your readers, you first need to have a firm grasp on it yourself.

"For folks who don't have a science background, it can sometimes be a little bit of a harder lift," Zaleski says. "I have a degree in English literature. I just have to make sure I talk to a ton of people, and that I understand the concept. I think over time it just gets easier, and you begin to realize what it is you're looking for."

Read the Studies—but Don't Get Too Caught Up

When you're looking for answers to the scientific inquiries you've found in a story idea, studies can really come in handy. In fact, you'd be surprised by the wide—and sometimes weird—array of questions scientists in university labs and research centers have asked, answered, and published papers on.

"The best stories strike a balance between being informative and relaying information from a study without bogging the reader down with details they don't need to know," Nazeer says. "So you need to be asking yourself how deep into a study you need to go in order to explain what the researchers found."

However, Nazeer points out, not all science reporting has to be based on studies. In fact, she says, that can be a bit of a pitfall.

"There are new studies coming out all the time, and you can end up on a kind of hamster wheel where you're not taking a step back to look at phenomena divorced from new research," she says. "The more interesting approach, as an editor and a reader, is looking at things that apply directly to everyday life and the hidden forces that shape how we think, behave, and interact."

Tell the Story

Ultimately, a story—whether it's about food, fashion, sports, or science, and regardless of where it's published—is still a story. Zaleski says his best ideas, many of which turn into science features, all start out as just a collection of interesting characters and scenes. His shark-hunter piece, for instance, was about marine biology and oceanic conservation, sure; but it was mostly a story about Mark Quartiano, "the Darth Vader of shark fishing."

"He was fascinating and entertaining, and he was the way to carry the reader through this broader story of ocean conservation," Zaleski says. "It always comes back to who the characters are. Is there an arc? Is there tension? Is there conflict? Then you're just building out the parts where the science comes in."

13 Pitching Errors: How Not to Pitch

Laura Helmuth

WRITING A GOOD pitch is really tough. Writing a bad one is easy. Editors see the same mistakes over and over again, even from good writers. Seven editors from a variety of publications participated in a roundtable discussion, in a series of group emails, about how *not* to pitch. I started the conversation off with questions, and then we talked among ourselves about our horror stories, pet peeves, and practical advice. Think of *The Open Notebook*'s Pitch Database as a lesson in how to make editors say "yes." Below, dear writers, is how to inadvertently make us say "no."

The editors who participated in the discussion:

David Corcoran, *New York Times*
Christine Dell'Amore, *National Geographic*
David Grimm, *Science*
Meg Guroff, *AARP the Magazine*
Laura Helmuth, *Smithsonian*
Robin Lloyd, *Scientific American*
Adam Rogers, *Wired*

LAURA: **Let's start with one logistical question: email or phone?**

ROBIN: Email pitches please.

CHRISTINE: Yes, definitely email pitches.

MEG: Email by a mile. The phone is intrusive. Email lets me see how you write, lets me forward your pitch to colleagues for consideration, and lets me ask follow-up questions or send a quick "no thanks" without getting dragged into a 20-minute conversation.

ADAM: Glad to see I'm not alone in preferring emailed pitches . . . and then emailed again, because I will admit to being the kind of editor who probably needs a prod to respond. Email is great for time-shifting, obviously; but the risk to the writer, especially one whom I don't know, is that unless the idea is a killer the email could fall below my horizon. And in truth, I often prefer a little preview-teaser email with just the logline of the idea—allowing me to say, "Sure, tell me more"—instead of the full pitch. It's weird how off-putting I have come to find the experience of clicking open an email only to have to wade through a two-graf anecdotal lede I'm not sure I'm going to care about. The best sales pitches, I think, start with a personal connection—and an opportunity for me to tell a writer, "Nope, I'm the wrong editor for this, you should email TK."

DAVID G.: Calling is a definite no-no in my book. I find it very unprofessional, and I especially hate when press officers do it (which is becoming a more common occurrence).

DAVID C.: Agreed on email pitching. I don't like cold calls, and I respond to them coldly. Freelancers seem to have gotten the message, because I get very few phone queries these days (and several a day by email). PR pitches are another matter, but let's not talk about them.

LAURA: **What's the most common mistake you see in pitches?**

ROBIN: Most common mistake—pitching a topic, rather than a story.

CHRISTINE: The most common mistake I see is freelancers who don't do their homework and read our website first—i.e., the majority of new pitches we get are for 10,000-word feature stories, like you'd see in *National Geographic* magazine, whereas we publish mostly 600-word-or-so news stories.

MEG: I'll agree with Christine that a lack of familiarity with the publication is the most common. Another is presenting a story as something you're dying to write, rather than as something our reader would be dying to read. Successful pitchers don't lead with their own desires or credentials. Instead, they focus on what's amazing about a story and how the story would fit into what the publication is trying to do.

ADAM: The most common mistake I see is a lack of familiarity with the magazine—pitches that are aimed as web articles, pitches on subjects we've covered (that don't advance the story), pitches for stories in a format or with an approach that *Wired* would never do. As an editor, I only want to feel loved, like the writer knows my true soul. Otherwise: no relationship.

DAVID G.:

- Not knowing the outlet (i.e., pitching us technology stories, which we almost never cover).
- Just forwarding a press release.
- Pitching the same stories everyone else is pitching (i.e., [studies from] *Science*, *Nature*, etc.).
- Pitching after the embargo has lifted.

DAVID C.: The single most annoying thing I see in first-time pitches is a lack of awareness of context. Why is this story suitable for the *Times*? What, if anything, have we said about it before? What makes it new? Why should a reader care about it now? This is basic homework every writer should do; if I don't see it, I'm most unlikely to read on. (I do try to respond to all emails that were individually sent to me.)

ADAM: Ooh, can I add one more tiny pet peeve? When a pitch consists solely of a writer saying, "Hey, did you see this? Might be worth a piece." And then copies in a URL. That is . . . not helpful to me. And it happens a lot.

LAURA: It's nice to see that many of these mistakes are universal—and so easy to avoid. Writers, use the search box. I often get pitches for stories we ran a few months earlier. Especially if it's a story that's been around a while, writers shouldn't assume they're the first ones to tell us about it. If they have some fresh angle, that's fine—but they have to spell out why their story is different.

David C., one of the mistakes I often see—and I bet everyone else does, too—is a writer pitching a story from Tuesday's Science Times.

I tend to get these pitches on Thursday. Do they think I don't read the *New York Times*? Or that I forgot the story within two days?

Adam, I love the relationship image. We want to be taken seriously, loved for who we really are!

As for how to tell a story from a subject, I hope any journalism professors out there invest some more time in teaching this distinction. Better to learn from a classroom, workshop, conference, or the *TON* website than to learn from years of failed pitches.

LAURA: **If you're comfortable revealing this, do you keep a blacklist? If so, what sort of pitch-related behavior does it take to get on that list?**

DAVID G.: I only put people on the blacklist after they write for me, usually because they either plagiarized or because their writing/reporting/attitude was so horrendous that I never wanted to work with them again.

ROBIN: It's not pitches that induce aversion to a writer; it's the quality of their writing and rewriting afterward, for me. Some writers send extremely weak pitches such as Adam described, but I'm still willing to work with the writer on those pitches at times, because I know they will do a good job in focusing/finding the story, and in writing and rewriting.

MEG: As for blacklists, there are a few writers whose tone-deaf pitching behavior demonstrates that I can't rely on them to represent me or the magazine appropriately. I had one writer pitching me periodically for years on the idea of profiling a particular 1970s rocker of whom she was enamored. No matter how frequently, gently, or baldly I declined this proposal—at first on the merits, and then because she was clearly not objective on the subject—it kept coming back, to the point where I was morbidly delighted to see it in my inbox. But of course I couldn't assign that story to her, or anything else for that matter. I just didn't trust her judgment.

ADAM: You guys have better blacklist stories than I do. I can't think of a writer I've decided never to work with based on pitches alone. It's always a function of what comes after the pitch—how good the work is, how easy the writer is to work with. I think I have a complicated cost-benefit algorithm involving the time it takes to produce a story versus the quality of the story, the quality of reporting versus the quality of writing versus the willingness to be edited, the quality of ideas (and their timeliness) versus the ability to execute them . . . I don't know. When they replace us all with AIs, the EditBot 6000 will be able to articulate all that much better. (But will it dream?)

Repeated mistakes, too, are a good way onto my blacklist. Rudeness or lack of cooperation with fact-checkers gets a yellow card and then a red card if repeated.

MEG: I've used writers whose pitches demonstrated they would not be able to pull off the story without a lot of help, but it had to be a truly brilliant, original idea, or a story that only they could tell. Doesn't happen often. Right now, I've got one would-be personal essayist with an unusual story who kindly agreed to be written about instead of hired.

LAURA: **What's the most horrible, ridiculous, epic-fail pitch you've ever gotten?**

LAURA: Mine involved one of the mistakes we covered already—forwarding a press release or news story—only the writer put a special twist on the mistake by not revealing that he was simply forwarding the guts of a newspaper story. He made it look like his own deeply reported pitch. The pitch was about animal cognition and it listed several then-recent examples of surprisingly smart behavior. You've probably heard of most of these studies—sheep that recognize individual faces, jays that stash food in different places depending on which other birds are watching, a New Caledonian "cow" named Betty that could bend a wire into a hook to extract food from a bottle.

Not knowing that the writer was merely forwarding a story from the *Guardian*, I asked some follow-up questions. I mentioned that I assumed he'd just mistyped the cow business and knew, of course, that it was a "crow" named Betty.

Next mistake: He didn't believe me. No, he said, his "source" confirmed that it was a cow. That's when I did a search and found the *Guardian* story and figured out that he'd plagiarized the whole pitch.

He was right, though: His source did confirm that it was a cow. On second reference, the story referred to her as a "bovine" that could bend wire into a hook—another reason publications should have editors with at least a smidge of science knowledge.

DAVID C.: Wow . . . I can't top that, and can't really think of the pitch from hell. I get a lot of mediocre pitches but nothing dramatically, howlingly awful. There was the *correction* from hell, which resulted from a freelancer's completely misunderstanding government data and confusing reported problems with actual injuries—a distinction the writer seemed incapable of grasping even after his sources explained to him that he'd gotten it wrong. This required me to write a whole new corrective article for the next week's section. Needless to say, he has never written for us again on anything remotely involving interpretation of data.

DAVID G.: As far as worst pitch, that would have to be a freelancer who pitched me a couple of years ago about an AIDS study. It was a very controversial study, promoting (if I remember correctly) an unusual therapy. Fortunately, I passed the pitch by our AIDS expert, Jon Cohen, who did some digging and found out that the freelancer's mother-in-law was an author on the paper. I confronted the writer about this, and he told me it wasn't a conflict of interest because he could be objective about the study. As we were going back and forth I noticed something else troubling: The freelancer himself was mentioned in the paper's acknowledgments. When I brought that up, I didn't hear from him again. Top that!

ROBIN: I can't remember an epic-fail pitch, but almost every pitch I get is fundamentally flawed—overly topic-driven (not a story), not tailored to our publication(s), full of structural problems, too long, too short, too publicity driven, or has factual errors in it. So I'm starting to rethink pitching and consider adopting these positions: (a) Pitching well is very hard to do, especially to multiple publications with their diverse audiences, tones, and themes; and (b) it is my job to work with writers who pitch to me to see if there is a good story in there for *SciAm* and to help guide them/us to it.

MEG: Ha, I defer to the son-in-law and the wire-bending cow. Most of my favorite horrid queries never get anywhere close to acceptance—they tend to involve writers going on for pages about themselves before mentioning a story idea. Once in a while I'll get a pitch from someone who wants to profile a celebrity, but wants my assurance that we'll take the story *before* even approaching the celebrity to request an interview! As if the mere fact that this person had heard of the celebrity were enough to merit an assignment.

ADAM: Worst pitch ever: I don't remember. I mean, the really bad pitches are easy. You just politely say no, and you never speak of them again. And the really great pitches are easy: You say, "Writer, here is money to do the thing you say you can do for me. Please don't suck."

The really hard ones are the pitches that *almost* get there. A pitch with a great idea embedded in really terrible writing just kills me. What do you do? Take a flyer on the writer? Make an inevitably ham-fisted attempt to buy the idea but assign it to someone else? No good options there. A well-written pitch about something that isn't right for *Wired*, or that we already did, often earns a "No, but please pitch again." A great idea in a pitch that won't get past our meeting process engages me— I work hard to develop pitches before my colleagues ever get a chance to evaluate them.

MEG: On a seasonal note, does anyone here respond favorably to Christmas cards from writers they've never hired? I'm on several of these lists and I think it makes the writer seem sort of lonely and bad at prioritizing. It's not a blacklisting offense, but not something that makes me want to hire the person, either.

CHRISTINE: To answer Laura's question, I don't keep a blacklist. I can't think of an epic-fail pitch, though there has been epic-fail *behavior*. We had a writer a few years ago who would pitch us a story then obsessively follow up. For one, he'd call each of us not long after emailing the pitch (we all sit in the same room, so each of our phones would ring in turn) and email us continually asking for an update. He became so annoying that eventually our managing editor had to remove him from our contributor roster. So I guess there is a limit to telling writers to be persistent!

I've never gotten a holiday card from a person we haven't hired as a writer, though I think it's nice to receive them from contributors. (That said, I think birth announcements, which we often get, are a bit weird!)

LAURA: Christine's anecdote about the writer who followed up obsessively immediately after sending in a pitch raises one question: How persistent is too persistent?

LAURA: We occasionally have a writer pitch a story to one editor, get turned down, then send the identical pitch to a second editor, get turned down again, and pitch again. That is too persistent. And a good way to get blacklisted—it's sneaky, and this business requires a lot of trust.

I usually reply to pitches within a week and don't mind getting a polite nudge if a week has gone by. If it's sooner than that, or not so polite, that's too persistent. (Exceptions for breaking news or a story with travel arrangements that need to be made immediately.)

ROBIN: Being reminded every two weeks on a pitch to which I've yet to respond doesn't bother me. More frequently than that is a bit annoying but sometimes it works too—at least for writers with whom I regularly work and who I know could probably sell a story elsewhere if I don't take it. I know they have to make sales, and that if I tarry, I am holding back their income and they are entitled to pitch elsewhere after some indeterminate amount of time. Relationships matter, again.

DAVID G.: I have a current freelancer, who, about two months ago began pitching three to four stories a day (I'm not making this up). He'd send them in a batch, or—even more frustratingly—send them one after another as soon as I rejected the previous one. I could tell from his pitch letters alone that he wasn't a good writer, and his machine-gun pitching was irritating the heck out of me. But . . . he was finding some good stories. So I let him keep pitching, and occasionally I took some of them. But to cut down on the pitching, I told him he couldn't pitch me more than one story a day. I still cringe a bit when I see his name in my inbox, but at least it's not as incessant as it was before.

MEG: It's a good idea to ask the editor when you should check back. Depending on the publication you're pitching, the proper interval could be daily, weekly, monthly. I appreciate a polite pester in the time frame I've suggested—it shows me the writer is eager and can follow directions.

CHRISTINE: I'd say anything beyond an email a week (unless, as Laura says, there's a time-sensitive element) is too much. Also, phone calls are not really preferred—it's better to respond to the person after I've had a chance to run the idea past the other editors.

ADAM: On persistence, I'm reminded of something *Atlantic* editor James Gibney said at a panel I was moderating: "There is a special place in hell for editors who don't call people back, and I am going there." I would say, if I haven't responded to your email in five working days, you should email me again and ask what's what. I will then make some kind of pathetic apology and dedicate some time to what you're pitching.

DAVID C.: I don't fault persistence, even to the border of rudeness; these writers are trying to make a living, whereas I have a relatively secure job. But as noted before, a writer who is clueless about our needs is unlikely to have the wherewithal to write a decent story for us.

LAURA: Does bad pitch hygiene get in the way of your relationships with freelancers?

LAURA: I mean relatively trivial things like someone not changing the subject line when they send a new pitch. Or answering a question but not appending the earlier email exchange, so I can't tell what the original question was. Or copying and pasting a pitch to some other magazine without changing the name of the magazine. (I get pitches all the time for stories that would be "a perfect fit for *National Geographic*.")

ADAM: "Pitch hygiene" is a great term for something that I've never been able to name. Like, I know I probably should just let it go when writers think they're being helpful when they embed a ton of links in a pitch (even though my build of Entourage unembeds them and leaves me with a document shot through with full-sized URLs, rendering it unreadable). I know I shouldn't hold it against them when a pitch shows up in three different font sizes, five different fonts, and a lot of boldface. And I know that it's an honest mistake when someone gets my name wrong, or the section I edit, or the name of my magazine.

But I'm going to come clean: This stuff makes me nuts. I am begging of you, dear writers: Make it easy for me to read your pitch. Let me introduce you to my friend, Plaintext. I think you would like each other.

Hey, that sending-pitches-to-multiple-editors thing is hilarious, huh? Six of our assigning editors sit within 30 feet of each other, with no walls between us. That double- (or triple- or quadruple-) teaming thing is something we notice, and are not kind to.

ROBIN: Thumbs up on pitch hygiene. It slows me down particularly when earlier exchanges regarding a pitch aren't appended. Then I have to go find the old email, and guess what . . . I won't. It's close to a kiss of death to your pitch. I have so many stories in my brain buffer daily—it's not that easy for me to remember your pitch/email from a few days ago without context.

MEG: Don't title your email "From [Your Name Here]." It indicates that you may be an idiot—all email programs tell you who the emails are from. I agree that links aren't great, but they're better than scads of attachments. I've had people send clips as PDFs, one PDF file per page. If you can't master this sort of thing, get your parents or children to help you.

DAVID C.: Thanks, all. I can't really add to those good thoughts, except to say multiple pitches are especially annoying—especially three or four in the same email, or the same week- or two-week period.

CHRISTINE: I'd say not really that much if the person has already proved herself/himself as a solid writer/reporter. I also tend to give people the benefit of the doubt at least once, probably because I've made similar mistakes myself as a freelancer!

ADAM: I do agree that I don't need to see a flurry of PDFs. When I'm ready to look at clips, I'll ask for them. Odds are if I'm thinking of working with a writer I'm going to Google him or her myself and root around, anyway.

LAURA: Do you have any advice for good writers (not the ones who misspell your name or pitch subjects rather than stories) about how they can make their pitches clearer, stronger, more efficient?

ROBIN: For writers with whom I have a relationship, sometimes they figure that means they can send short, two-sentence pitches all the time. I'd still prefer a longer pitch of three or so brief paragraphs. Some pitches go on for four or five lengthy paragraphs, or longer—that is too long for my purposes.

DAVID G.: Final advice:

- Always write "pitch" or "query" in the subject line, so I know it's not a press release (and so I don't automatically delete it).
- If you're not sure what to pitch, write me and ask me what sorts of stories I'm looking for.
- Don't pitch the big stories from *Science* and *Nature*. That's what everyone else pitches. Find me the cool, under-the-radar stories that will become exclusives.
- Since you're pitching me a web story, always mention if there's multimedia. Sometimes that can put a mediocre story over the top.

ROBIN: Oh yes. I want to underscore these points of David's:

- Write "pitch" in subject line please.
- Don't pitch from *Science*, *Nature*, *PLOS*, or *PNAS*—I've got those covered.
- If you're not sure what to pitch, email me and I will send back a standard "what I seek" email that I have prepared for such purposes.

MEG: Advice for good writers: Trust your story. Don't start your pitch with who you are or who we know in common. Grab me with a lead-in that shows what a fantastic idea you've got and what a fantastic writer you are. Then you can briefly state the qualifications that make you perfect for the assignment, including anyone I know who can vouch for you, if there's anyone.

Don't offer to provide photographs unless they are rare historical images. We're a glossy magazine that works with top photographers, and unless you regularly shoot for *National Geographic*, your photos are not going to cut it. The offer makes you look like you don't understand what we're doing.

Be nice. Life is short and editors are human beings. I would much rather work harder to coach someone who is open-minded and pleasant than invite a known jerk into my life, no matter what their copy looks like.

CHRISTINE: In our writing guidelines, we ask freelancers to send us a potential headline and 135-character summary along with their pitch. This helps zero in on the news quickly, especially if their pitch is rambling or unclear.

Echoing my first comment in this thread, I can't stress enough that the person shows some knowledge of the publication in their pitch. If they're pitching a news story about a new species of Indonesian frog, mention that we covered another species in the genus in March 2010 and this would be a great follow-up, etc.

ADAM: Meg, I disagree with you about starting a pitch with a zinger of a lede. Even if it's great, that's two paragraphs I have to slog through before I know what the story is about—assuming it's a magazine-y anecdotal thing. I'd much rather my first round with writers—whether I know them or not—be less formal to start, on the order of, "Hey, I have a story about TK. It's important for these reasons. . . . Would you maybe be interested?" I may be wrong about this, but I also prefer that kind of informal exchange as a way to assess writing skills. It's all an audition, right?

Advice for good pitches? Don't write a pitch longer than the story you'd be assigned. Our front-of-book section stories rarely go longer than 300 words. Know what section you're pitching, and maybe even what kind of item. Using the terminology of my magazine has the double benefit of making my life easier by saving me from having to think about something, and also proving that you know whom you're pitching. Be clear and concise—there'll be time for stylistic shenanigans later.

14 Five Ways to Sink a Pitch

Siri Carpenter

SELLING EDITORS ON pitches is always a gamble. To maximize your chances of success, avoid these common mistakes.

1   **Pitching a story the publication or its main competitor just ran.** If you're pitching a story that's meaningfully different from others on the subject, show why the difference matters.

2   **Pitching a story unlike anything the publication ever publishes.** Do your homework to understand what types of stories are realistic for the outlet you're pitching, both in substance and in format.

3   **Pitching a topic, not a story.** A story is driven by some central argument or question that you will set out to address—that is, it makes a point. Another way to think of it: Topics are nouns; stories are verbs.

4   **Pitching a series of questions you'd like to ask, but no answers.** You can't know everything when you pitch. But you should have a decent sense of what answers your story will offer readers.

5   **Sending a pitch full of typos or other errors.** Everyone makes mistakes. But editors are looking for clues that you can deliver clean, accurate copy. Don't give them reason to worry that you're sloppy.

15 What Makes a Good Pitch?
Annotations from the *TON* Pitch Database

Roxanne Khamsi

IF YOU WERE to believe the movies, you would think that the first time a journalist sits down to write is after they get their assignment, often when most of the reporting is done. In reality, the writing process almost always begins far earlier. It begins with the pitch. This is the true starting point—and getting it right means that you can land the assignment. Given the high stakes, it's well worth investing time in the pitch so that it contains all of the key elements to entice an editor to say yes. If you have an idea for a long magazine feature, for example, you'll want to get on the phone with possible sources and do some prereporting so that your pitch gives a taste of the characters and complexities involved in the story you hope to write.

A good pitch will convey the crux of the story you envision and its timeliness, often with a description about an event or a new study that can act as a news hook. A *great* pitch will do all those things, and it will also stir the imagination with vivid imagery and introduce compelling characters involved in an evolving story with high-stakes outcomes. The four pitches annotated here, selected from *The Open Notebook*'s Pitch Database, all use vibrant language to underscore the urgency of the stories they seek to tell, and they offer inspiring templates to follow, whether you are pitching a short news article or an epic magazine feature.

The Story

"How Illegal Mining Is Threatening Imperiled Lemurs" by Paul Tullis

For *National Geographic*, published March 6, 2019

The Pitch

In a densely wooded area of northeast Madagascar last fall, someone discovered a blue gemstone: A sapphire. *[[This first sentence uses descrip-*

tive language to set a scene and transport the reader to a different landscape, while simultaneously hinting that a recent discovery of a precious gem may precipitate big changes. The following sentences raise the stakes even further by conveying just how grave the situation is for lemurs as a result. Building this tension right off the bat grabs an editor's attention.]] That initiated a rush of people into an ostensibly protected wildlife corridor intended to protect lemurs; more than 90% of lemur species are on IUCN's red list due largely to habitat loss and fragmentation. This was at least the third instance in recent years of the discovery of sapphires leading to thousands of Malagasy destroying lemur habitat in search of riches. Official estimates put the number of people at the site near Ambatondrazaka last fall at 4,500, forming a large tented community where the cries of critically endangered indri lemurs could be heard among the sounds of chopping trees and digging gravel. *[[When you are crafting a pitch, you want to demonstrate that you have the writing chops to deliver the kind of article the publication would publish. This pitch does that by not just stating the facts but also embedding them alongside visceral details about the sights and sounds in this remote area. This is exactly the kind of captivating storytelling that one would expect in National Geographic, which is where this story was pitched and accepted.]]* Women in jelly sandals trudged for 12 hours through the muddy jungle with their young children in tow, some to dig for jewels, others to cook and make coffee for the miners. Local traders—mostly Sri Lankan—believed the stones rivaled Burmese sapphires, which are considered the best on the market and demand the highest prices, though few of the Malagasy digging in the mud found much of any value. After a few months, gendarmes who had been complacently observing the scene for weeks suddenly decided to shut down the mine.

Madagascar's government is weak, and lacks the capacity to enforce wildlife and forest protections across the country. *[[After the first paragraph sets the scene, this paragraph gives the larger context for the story— much like a nut graf does in a news piece. This helps give the editor a broader sense of the urgency of the story and why it matters.]]* A land deal the government negotiated with a foreign corporation several years ago led to protests resulting in a coup, with the international community not recognizing the government that followed. All US aid except humanitarian was cut off, and the instability led to a rush for illegal deforestation of rosewood. The country's wet, jungly east, where forest cover had been expanding, saw tree cover loss of 20% or more in many areas in the years after the coup, according to Global Forest Watch. *[[A striking number such as this, from a reputable watchdog group, lends weight to the story. If you have facts and figures from independent groups that help emphasize the importance of your story, use them.]]* The current president was a member of the coup administration and corruption is rampant. *[[Throughout this pitch, it's clear that the writer has done his homework. This gives the editors confidence to assign the story because they can trust that the writer has cultivated expertise in this area.]]* One miner near Ambatrondrazaka told a visitor he was there to find a gem so he could pay a bribe his university's administration had demanded to issue the degree he had earned. "The government is engaged in cronyism in that very region [around Ambatondrazaka]," said Steig Johnson, a professor at the University of Calgary who spends several months a year in Madagascar studying lemurs. Not far away, in Didy, the president's son is involved with Sri Lankan traders in illegal sapphire mining, according to Jonah Ratsimbazafy of MICET, a conservation group. *[[Here the pitch makes it known that this isn't just an isolated trend; this practice of sapphire mining and trading is affecting all parts of the country—even reaching the families of those in leadership. This makes it clear that this is a big story.]]* "It's common now; people don't care if it's a strict protected area. They just go in massive numbers to extract stones and nobody can stop them," he said. "It's linked with corruption and the laws are not really enforced. Kids quit school because they want to look for sapphires, the water is made dirty, there's insecurity. There are rich people who push it and protect it, that's why you cannot stop it. It's a mess." *[[Not every pitch needs quotes, but pitches for magazine features like this one are best when they include them, because quotes show the editor that you have already made contact with relevant sources and that the information in the quotes themselves bolsters the case for your story.]]*

The political instability has weakened Madagascar's economy, forcing many residents to move to new areas in search of opportunity, often in illegal sapphire mining. Migrants often don't respect local cultural taboos, including those against killing lemurs, and bushmeat hunting can now be added to the list of threats to the lemurs. *[[This pitch continues to raise the stakes. We already know lemurs are under threat because of habitat loss, and now we learn they are also threatened by bushmeat hunting. A strong pitch will add layers of urgency to the story with every successive paragraph.]]* "Those from the south respect taboos in their region but when they move to other regions they don't respect the taboos of the other tribe," said Ratsimbazafy.

But MICET and international organizations are working to save lemurs by helping local communities see value in their surroundings other than through resource extraction. *[[The editor learns in this paragraph that there is more than just doom and gloom—there is hope. All is not lost. This potential for a solution appeals to editors.]]* Ed Louis, a wildlife biologist with Omaha Zoo, started the Madagascar Biodiversity Project 10 years ago and currently runs the biggest reforestation program in Madagascar. It involves engaging local communities on a number of levels, including as employees making a lot more than the $0.55 per day average in the country. *[[Here it becomes all the more evident that this is not just a story about conservation. It's also about economic factors shaping the landscape of the country, and people struggling just to get by. The information here makes it clear that this is a complex issue that deserves a long feature, not just a news brief. If you are pitching a feature, make sure there are layers to the story like this.]]* Women who help to plant trees are paid in credits they can use to buy a sewing machine so they can develop income independent of men in the community. "It's not going in and saying, 'Stop destroying forests,'" Louis told me. "If the people aren't on board, all the species we're doing research on will not be there much longer." MBP helps farmers get deeds to their land and places the documents in several offices around the country; in exchange the farmers commit to setting aside a portion of their property for lemur habitat. Working with another NGO, MBP developed a coloring book for kids that discourages the hunting of lemurs; they distributed the spiral notebook and 5 crayons to 5 sites with critically-endangered lemurs and saw hunting come pretty much to a halt as kids convinced their parents to quit the practice. "Whether you have clean water, forest, disease, medicine—it's all connected," Louis said. "If we provide a rocket stove, that family doesn't have to cut wood. If the water is clean, a guy doesn't get sick, and he can work to feed his family." The approach is seeing some success, with counts of some critically endangered species up significantly since Louis began his work. *[[This paragraph is crucial: Not only does it introduce a compelling character who has worked for a decade on this issue, it also ends by saying that the program has already achieved some success. This tells the editor it is not just a good-hearted man's pipe dream—it is a plan that has started to work.]]*

I'd like to travel to Madagascar for at least a week to report on sapphire mining, government intransigence in doing anything about it, and conservation efforts to save lemurs despite it. According to my sources *[[Here the writer shows that he has access—perhaps giving him an exclusive to this particular story.]]* it's very likely that sapphire mining will be ongoing at least on a small scale at or near the Didy or Ambatronrazaka sites (or if not there, somewhere else nearby, and if not there the destruction will be visible and former miners will be easy to find). *[[If you hope to travel for a story to do on-the-ground reporting, make sure to lay out a plan, as this pitch does. This paragraph specifies how long the trip would be and what kinds of sources the writer would meet with. It also tells the editor that there is an opportunity to actually observe the mining that is at the heart of this story.]]*

In any case, thanks for considering. I look forward to your response.

Thanks,

Paul

The Story

"Meet the Computer Scientist You Should Thank for Your Smartphone's Weather App" by Sarah Witman

For *Smithsonian*, published June 16, 2017

The Pitch

Pitch: The "hidden figures" who changed weather forecasting *[[Adding a title like this to a pitch automatically places it head and shoulders above the many pitches that lack this feature. A title gives an editor a quick capsule summary of the idea, and the word "pitch" also alerts them outright that this is something you would like to write for the publication.]]*

Hi [editor],

I'm a freelance science writer based in Madison, Wisconsin (we met briefly at the last NASW meeting in San Antonio). *[[If you met an editor—even in passing—and you are trying to break into a publication for the first time, remind them like this that you crossed paths. It automatically creates a connection with the editor and engages them.]]* Over the past few years, I've written about cybersecurity in the food industry for *Quartz*, HIV in Malawi for the *Pacific Standard*, spaceflight's impact on the human body for *Popular Science*, and more. *[[This sentence establishes the writer as a person with experience who has a track record of reporting on science. If you're just breaking into science writing and don't have a ton of clips, don't fret: It's not a necessary sentence to include when you're first getting started. But as soon as you have some articles to point to, add them in the pitch—either at the start or in the final paragraph. You don't want to bombard the editor with a résumé-length list of accomplishments in the text of the pitch—just highlight the most relevant few pieces you've written (and perhaps include links to the stories, as this writer did).]]*

I'm working on a story that I'd like to run by you. It would be a profile of Klara von Neumann and the other "hidden figures" who changed weather forecasting as we know it. *[[This sentence accomplishes two things quite brilliantly: First of all, it gives the editor a distillation of the story to grab them right off the bat. Secondly, it references the well-known book (and movie hit) Hidden Figures, about a group of Black women who did key calculations that helped NASA launch astronauts into space. This sentence teases the idea to the editor that there is another important story about underrecognized women who made massively important contributions to a different area of science without getting their due.]]*

Generally speaking, modern meteorologists forecast the weather using a technique called numerical weather prediction: They input current conditions into computer models, which estimate what the weather will do in the future. *[[These first two sentences of this paragraph give the editor a succinct description of the way weather is currently forecasted. It's important not to let the details of the science you want to write about overwhelm your pitch, but remember that the editor might not know anything about the subject—so explaining the central scientific element can help.]]* This technique can be traced back to a single paper, published in 1950. A team of scientist—top minds in meteorology and computing—used one of the first computers, ENIAC, developed during World War II, to produce the first numerical weather prediction. Or, as the Weather Bureau put it, "These men had made the first successful barotropic forecast on a computer." *[[Here, the pitch makes it clear that the writer has taken the time to get to know the history of this area. This makes the editor more convinced that the writer has what it takes to deliver a stellar article, and thus more confident to say yes to the story.]]* Except it wasn't just men. *[[A short declarative sentence like this helps spark interest.]]* Klara von Neumann, who was married to one of the paper's co-authors, acclaimed mathematician John von Neumann, was instrumental to the project. *[[We've met Klara von Neumann earlier in this pitch, but here the writer expands on what we know about her.]]* During World War II, Klara worked in Princeton's Office of Population Research, calculating population trends. So when John started working on ENIAC after the war, she was well-equipped to become "one of the first 'coders,' a new occupation which is quite widespread today." When the team convened in early 1950, it was Klara who "initiated them into the ways of the ENIAC and its peripheral card-processing machines" and "checked the final program." *[[Here, the writer hyperlinks to quotes supporting the case that Klara von Neumann was a bona fide coder behind the early weather-prediction computer models. By referencing supporting materials, the pitch makes it clear this is not just a theory put forth by the writer—it's verified by other people.]]*

Yet, Klara is not mentioned in almost any papers or books about the history of weather forecasting. *[[Here the writer is adding complexity to the story: Not only were the women involved, they were also overlooked. The sentence also signals to the editor that the journalist wants to report a story that readers have likely never heard before. Editors are always looking for fresh and untold stories of consequence.]]* In the 1950 paper, she is simply listed in the acknowledgements, when the same work today would certainly earn her co-authorship. She is literally a footnote in history. The "staff of the computing laboratory of the Ballistic Research Laboratories" (which, at that time, was mostly women) also received an acknowledgement, though not by name, "for help in coding the problem for the ENIAC and for running the computations." And Margaret Smagorinsky, Norma

Gilbarg, and Ellen-Kristine Eliassen were not thanked for their "hundreds of hours of hand-computing" in trial runs before the ENIAC was up and running. *[[The pitch continues to evolve and expand the story here. It's not just Klara von Neumann who has been denied acknowledgement—it's several other women, too. This makes the story seem bigger and underscores why it deserves more space than just a short news piece.]]*

I would like this piece to focus on Klara's contributions to this project, as described above, in the context of the many forgotten contributions of women in the early days of computing, which have only recently started to become well-known. *[[Here the writer reminds the editor about the timeliness of the story and how it could be framed in the context of other women finally receiving recognition for their work.]]* I've interviewed Dr. John Knox, who teaches his students at the University of Georgia about Klara's role in meteorology history, and Peter Smagorinsky, whose deceased parents both worked on the 1950 project. *[[This sentence shows that the writer is invested in this story and has already taken the initiative to interview sources.]]*

Please let me know if this interests you. Looking forward to hearing your thoughts!

Best,

Sarah Witman

The Story

"Thunderstorm-Triggered Asthma Attacks Put under the Microscope in Australia" by Katherine Kornei

For *Science*, published January 25, 2018

The Pitch

After a thunderstorm danced over Melbourne, Australia in November 2016, thousands of people were hospitalized and nine died. *[[Pay attention to the verbs in your pitches and make them as active and engaging as possible. Here, for example, the thunderstorm doesn't just occur or arrive—it dances! The writer combines this dynamic image with the hard-hitting fact that nine people died as a result of the event. This tells the editor that the storm was massively consequential—and makes them wonder how a thunderstorm could be so widely fatal.]]* But lightning strikes weren't to blame. Instead, deadly asthma attacks were responsible for the casualties, which included people with no prior personal history of the disease. Welcome to the new world of "thunderstorm asthma." *[[The paragraph reveals that there is a threat unknown to readers that could affect their health. It justifies the need for the story. The phrase "thunderstorm asthma" also serves as a useful two-word summary of what this piece is all about. If your story relates to a phenomenon with a catchy label like this, use it (and use it early on).]]*

This phenomenon, first identified in the 1980s, isn't well known, even among meteorologists, says Andrew Grundstein, a climate scientist at the University of Georgia in Athens. *[[Not only do readers not know about the subject—many experts in the field don't either! This sentence hints that the story is unique and shows that the writer has connected with a rare scientist who is already in the know about the phenomenon. It also gives useful context about the origin story of this scientific area, which hints that there might be interesting history to delve into and that this weather event isn't just a flash in the pan.]]* Data are scarce, but thunderstorm asthma is believed to be triggered by a "perfect storm" of heavy rainfall and electrical activity rupturing pollen grains and then gusty, down-drafting winds associated with thunderstorms sending those micron-sized grains flying deep into lung tissue. *[[Here the pitch gives the distilled summary of the science behind the story. It's perfectly short and sweet, and it gets the job done. The verbs here—"triggered," "rupturing," and "flying"—make what could have been a dry description of the science come alive.]]* Melbourne, Australia seems to be a hot spot for thunderstorm asthma: it's surrounded by grassland, is prone to thunderstorms, and is home to a large population (over 4 million people).

Now, researchers led by Grundstein have examined 7 thunderstorm asthma events in Melbourne from the 1980s through the present day. *[[Nothing makes an editor happier than a solid news peg. The first sentence of this paragraph offers just that (and more details are given about the related study in the next paragraph). Be sure to include a sentence like this that highlights the timeliness of the story. The editor will want to know why this story needs to be reported and published now, as opposed to a year ago or a year from now.]]* With the goal of predicting what weather properties are most likely to trigger thunderstorm asthma—which could help hospitals better prepare for a deluge of patients—they retrieved archival weather records from each event. They found that down-drafting winds were critical to spreading the broken pollen grains; the presence of cold fronts was also linked to thunderstorm asthma events. *[[This sentence tells the editor the results of the study, and it does so clearly. That's important! Don't make the mistake of saying a new study was conducted without also relaying the results. And when you explain the results don't just parrot the scientific jargon of the paper. Take the time to digest the real meaning of the findings and summarize them in language the editor will easily understand.]]*

These new results were presented this week at the American Meteorological Society conference in Austin, Texas (abstract link below), and the 2016 Melbourne case was published last year in the *Journal of Applied Meteorology and Climatology* (manuscript attached). *[[If there's a study at the heart of the story you are proposing, editors will want to know if the study is published or perhaps presented at a meeting. Follow the example here of noting this to the editor. If it's as-yet-unpublished work, make that clear.]]* I've chatted on the phone with Grundstein, and he's happy to be interviewed further. *[[This writer is on the ball! She has already spoken with the lead scientist behind the project. This signals to the editor that the story is in motion and the writer is on it.]]*

This subject is particularly timely because the Australian state of Victoria—home to Melbourne—recently started using a thunderstorm asthma warning system. Grundstein believes he and his team can help improve the algorithm to more accurately predict when these potentially deadly events will occur. *[[With each paragraph, this pitch makes an even stronger case that the time is ripe for this story. Here we learn that a warning system is already being deployed, and that the expert believes it needs fine-tuning.]]*

The incident in Melbourne made headlines in 2016 (links below), but I haven't found a story that really digs into the science of what weather conditions create this deadly phenomenon, which I think would be of interest to *Science*'s readers. *[[Editors really want to know what kind of coverage a story has received and how your story will be substantially different and therefore warranted. They don't want a story that is just a small iteration on previous reporting. Here, the writer is showing how earlier stories were simply delivering the news about the events but not digging into the ac-*

tual causes behind them. She also demonstrates she's done her homework by looking around for stories like the one she wants to write (and finding none). Stating that explicitly to the editor can save them time from having to look around themselves for related coverage (although a good editor will double-check themselves anyway).]]

How about a story for the magazine?

Thanks,

Kathy

The Story

"Why Are There So Few Women Mathematicians?" by Jane C. Hu

For the *Atlantic*, published November 4, 2016

The Pitch

Hi Ross,

As soon as mathematician Chad Topaz ripped the plastic off his copy of the American Mathematical Society's magazine *Notices*, he was disappointed. *[[Wow, this opening sentence has it all! It paints a scene and describes an action. It also introduces a character and immediately conveys that person's emotion during an event. The mathematician didn't simply handle the magazine, he ripped the plastic off of it. This verb conveys urgency. If you pay attention and carefully select verbs this way, it can do wonders for your pitch. Because of this verb we know that the mathematician was excited for the magazine, and we are immediately curious as to why he was disappointed.]]* Staring back at him were the faces of 13 fellow mathematicians—all of them men, and the majority of them white. *[[This is the second part of the one-two punch of the opener. Now we understand why the mathematician was disappointed, and we understand the problem at hand.]]*

Topaz, a professor at Macalester College, knew that his field had a gender problem. In mathematics, just 15% of tenure-track positions are held by women. *[[This context is useful to the editor and validates that the problem mentioned in the first paragraph is really emblematic of a wider imbalance in the field. The author has also noted the scientist's institutional affiliation here. That is useful for the editor to know, too.]]* Loads of recent research has shed light on how women are underrepresented in top labs and university research faculty—but Topaz was determined to understand the finer-grained details of what could be driving these disparities. So Topaz and colleague Shilad Sen decided to look at a new metric of academic success: the editorial boards of academic journals. *[[The pitch is evolving and adding layers to the story. The character we met in the first paragraph isn't just any mathematician—he is a mathematician invested in solving the problem of gender disparity, and he is working with a colleague to unravel the root causes of this issue. Our curiosity is heightened here because we want to find out what they discovered.]]*

According to Topaz and Sen's analysis, just under 9% of math journal editorial positions are held by women. Their research, published in *PLOS One*, analyzed the editorial boards of 435 math journals. *[[These two sentences offer the editor crucial information: the scope and results of the study, as well as where it was published.]]* I'd like to propose a piece on how why this duo undertook their analysis, and what the gender imbalance in academic journal editorial boards tells us about the under representation of women in science. *[[The writer describes her vision for the story here, and it is not pure conjecture—the pitch has already stated facts to support this approach.]]* Editorial positions are an especially important role for scientists; editors are the gatekeepers of research, deciding which papers get published, which, in turn, sets the tone for the field about which areas of research are worthy of study. Furthermore, serving on editorial boards is an important networking and professional development opportunity for researchers. Being left out of these opportunities can affect female researchers' careers. *[[Here the writer is connecting the dots for the commissioning editor. Don't just expect the editor to understand the significance of a study's findings: Spell it out like the writer does in this pitch.]]*

I'm in touch with the researchers, and I've previously written on studies investigating gender disparity in the sciences, as well as on the culture of academia in general (including a piece for the *Atlantic* earlier this year). *[[The writer tells the editor that she has already written for this publication. That's huge! If you have already reported a piece for the same outlet you are pitching, make sure to let them know in the pitch, because it helps establish your reputation as someone who can deliver the kind of copy they seek. She also notes here that she has been in touch with the researchers behind the study that she wants to cover. This shows her initiative and highlights that she has already gained access to central characters in the piece. The writer also makes it known that she has covered other news in this area before, giving the editor confidence that she has the expertise to ask the right questions and find the right angle for this particular story.]]* You can find links to my other work at janehu.net. Happy to elaborate on any of this or any answer any questions you may have—thanks for considering!

Jane

SOME WRITERS SAY that crafting a winning pitch is the hardest part of the writing process. You might have to vividly describe something happening in a faraway land that you haven't yet had the ability to travel to, or you might not yet know the full arc of the story because the events are still unfolding. Even faced with such circumstances, though, you can still write a great pitch by choosing hardworking verbs to paint a gripping scene and distilling the key scientific advances into direct and jargon-free language. A lot of writers worry about how long their pitch should be.

It's important to remember that a great story idea can be teased in a single sentence. The news peg, urgency, scope, and exclusivity can all be packed in tight if you need to. In a 2015 piece for the *New Yorker*, writer John McPhee describes how he went to his editor at the publication decades earlier floating a diffuse idea of writing about oranges. "I had asked Mr. Shawn if he thought oranges would be a good subject for a piece of nonfiction writing," McPhee recalls. "In his soft, ferric voice, he said, 'Oh.' After a pause, he said, 'Oh, yes.'" McPhee's massive story on the subject ran in the magazine in 1966, and later became a book.

That exceptional level of creative license is rare, of course. Staff writers generally have more leeway to pitch story ideas before fleshing out the details, compared with freelancers. So a pitch needs to be long enough to frame the story and make your qualifications for reporting it clear. If you are pitching a long feature article, each paragraph should add intriguing complexities to the story.

Every pitch is different. But all great pitches convey the excitement and expertise of the writer. If you invest the effort into pitching a story, chances are strong that an editor will invest in you to write it.

16 A Conversation with Kathryn Schulz on "The Really Big One"

Michelle Nijhuis

KATHRYN SCHULZ'S *NEW YORKER* story "The Really Big One," published on July 13, 2015, opens in Japan, moments before the 2011 Tohuku earthquake. American seismologist Chris Goldfinger, who is attending an international conference in the city of Kashiwa, feels the room begin to shake. At first, he and his colleagues laugh dismissively, assuming that they're experiencing one of Japan's frequent small quakes. But the shaking goes on, and on, and on, and soon, everyone in the room realizes that something very big is happening.

As the world now knows, the quake turned out to be the largest in the country's recorded history, and the subsequent tsunami devastated northeastern Japan. For Goldfinger, a professor at Oregon State University, the Tohuku earthquake was a frightening preview of another disaster—one much closer to home.

Goldfinger studies the rising tension in the Cascadia subduction zone, off the coast of Northern California, Oregon, and Washington. He and his colleagues estimate that the chance of a "big one"—a quake of magnitude 8.0 or higher—happening in the zone within the next 50 years is about one in three. As Schulz vividly describes, such a quake would have massive consequences, inundating much of the coastal Pacific Northwest and causing enormous damage in Portland and Seattle.

When her story appeared in early July, it immediately went viral. As Schulz reported in a later blog post, readers described it as not just "terrifying" but also "truly terrifying," "incredibly terrifying," "horrifying," and "scary as fuck." The line that got the most attention was a quote from FEMA official Kenneth Murphy: "Our operating assumption is that everything west of Interstate 5 will be toast."

Schulz, a staff writer at the *New Yorker*, first heard about the inevitable rupture of the Cascadia subduction zone from a relative in Portland, Oregon. "I honestly couldn't tell you whether my very first reaction was horror for my beloved Pacific Northwest, or a kind of writerly thrill of

'Oh my God, that's an amazing story,'" she says. Here, she tells the story behind the story, and the story after the story.

MICHELLE NIJHUIS: When did you first hear about the Cascadia subduction zone?

KATHRYN SCHULZ: Part of what immediately made this a compelling story to me was that even though I lived in Portland for several years some time back, I have a bunch of family out here, I have really good friends out here, and I spend my summers here, I only heard about the Cascadia subduction zone last year. It was, as you might imagine, kind of a shock to the system. I mean, I'm a major consumer of news. I read the paper, I listen to NPR, I actually know people in politics and policy. I thought, "If I haven't heard about this story yet, something is a little amiss."

Other than the initial shock of "Wow, this is possible," what made you think, "The time is right for me to write about this"?

I don't think it was so much a "time is right" feeling as just an "Oh, obviously." As a writer, you're only so often presented with a story that you immediately know is great, and that you immediately know you want to write. I honestly couldn't tell you whether my very first reaction was horror for my beloved Pacific Northwest, or a kind of writerly thrill of "Oh my God, that's an amazing story."

How did you know you wanted to write it?

Not because I have any type of particular skill set or aptitude—a lot of people could have written it. I just had the happy accident of being in its path. And I think it helped that I really love the region, I'm really familiar with it, and I already spend a lot of time out here, so reporting-wise it was a bit of a no-brainer. But I also love science in general and geology specifically, and I'm really interested in the relationship between humans and the rest of the natural world. So in that sense it was obviously a story that I was going to want to write.

What were the first research steps that you took?

The very first person that I got in touch with, on a tip from the person who first told me about the Cascadia subduction zone, was Carmen Merlo, who is the head of the Portland Bureau of Emergency Management. And then I did what reporters do, which is that I asked her who else I should talk to. She was immensely helpful in describing what's going on in the city of Portland, but she's also been in emergency management in the Northwest for a long time, so she was able to hand me a vast list of names.

When and how did you talk to your editors about the story?

Very early on. I brought the story to the *New Yorker* as a freelancer, and in the middle of my working on it they hired me as a staff writer. So the story bridged those two different relationships with them. But I pitched the magazine as soon as I'd done enough background research to get a feel for the accuracy of what I had been told, and the scope of the problem.

Chris Goldfinger is a really compelling character. When did you know that you were going to use him as the central character in the story?

Very late in the game! I mean, Chris was always going to be in the article. He's one of the leading scientists on this subject, and he's the person who gave us the timeline for what we should expect from Cascadia. He was incredibly helpful to the article, and in every incarnation of it he was present. I cited his work, I talked about his numbers, and I think his lab showed up in most drafts. But the story came out in July, and until about May the draft of this piece looked very, very different. And among the starkest differences was that there was almost no Chris Goldfinger in it. It did not begin the way that it currently begins, it did not have him as sort of the central character. He was in it, his work was in it, but his appearance was not much stronger than a cameo.

How did it originally start?

The early version of this piece started in the town of Seaside, Oregon. So the whole thing flip-flopped. Now Seaside has a cameo, in the final version, and Chris plays the leading role. Originally Seaside was almost the main character—instead of being built around a person, the piece was built around a place. And there were some really compelling reasons to do it that way. Seaside, with one possible exception, is the town that stands to lose the most when this thing hits. Something like 90 percent of the town is inside the tsunami inundation zone. It is really a very bad situation. So that was compelling. There were a couple of people there who were compelling and did a good job of speaking for the city. And it gave me a visual: I could hang out in Seaside, right there by the ocean, and in a different way than I ultimately did I could pan out beyond the shoreline, out to sea, and drop into the ocean to show people where these plates were meeting.

So it was a very different opening. It was quieter in some ways —I think I was aiming for a kind of quiet menace, but it might have also been just sort of sleepy.

What were you going for by changing to a character-based opening?

You know, I set the piece aside for a long time, because after I was hired as a staff writer, I had these more time-sensitive articles I had to get done for the magazine. And when I went back to it, I could see where I was coming from. I understood why Seaside had been tempting, and why a place-based version had been tempting. I thought the writing was kind of pretty. But then I thought, "Oh my God, Kathryn, this is a mega natural disaster that we're talking about—give the people what they want!" I realized that it was not the moment to be lyrical and sleepy and slowly build the subject. It was the moment to really put people inside this experience, and let them see it unfold, see what it's like. And place can be powerful, but sometimes people are more powerful. I'm sometimes resistant to the obvious move, but I had a moment when I was looking at my draft and I thought, "Schulz, there's a reason people use those obvious moves. They work!"

And since you're writing about something that hasn't happened yet, it gave you a way to start with a version of the real event.

That is exactly right. When you first think about this story, you might think, "Oh, what could be easier? People love natural disasters, this thing's going to write itself." But what I realized almost right away is that it's very difficult to tell the story of something that hasn't happened yet. When it isn't fact, when it isn't what has happened or what is happening—when that particular tool is taken out of the tool chest, you can get in a lot of trouble, or have a lot of difficulty. I didn't want to do the, you know, "It's 9:45 a.m. in downtown Seattle, the commuters are . . ." I'm not wild about those kinds of thought experiments. I've seen them done really well, but they can also be very hackneyed. So I made the closest move I could, which was to go to an analogous situation. And though it was a little bit of a bait and switch, it actually is the closest example we have in recent history of what's going to happen here, and it's a terrifying one—because that place is much better prepared than ours.

For me, a big part of the scariness of the story comes from your decision to use the future tense rather than the conditional: "The way it will arrive . . ." and "The destruction will begin . . ." It's a small thing, but I think that grammatical choice has a huge emotional impact. How did you make that decision?

First of all, the fundamental truth about the Cascadia situation is that it is a when, not an if. So in a funny way, the science made the decision for me. Now, inevitably there are micro-conditionals: Am I going to get hit by a brick from across the street? There will be huge differences if it strikes in winter versus summer, or if it strikes in daylight versus nighttime. There are a lot of unknowns. But the fundamental fact of its happening is a known, and you know, grammar is your friend. Strangely, it creates its own kind of plot, and you can use it.

So it was a very deliberate decision to not use the conditional. And for all the limitations that come with telling a story that hasn't happened, as we were just discussing, it's kind of a rare, weird pleasure as a journalist to write in the future tense. There's something about the future tense that has this momentum, this feeling of imminence and suspense.

I wanted to ask you about the hand metaphor that you use—the way you repeatedly ask readers to use their hands to mimic the action of the plates in the subduction zone. It's so effective. Was that the way the movement was explained to you, or was that something you came up with in the process of writing this story?

It was not explained to me that way, but I don't remember any incarnation of this story where I did not explain it that way. I think I myself must have sat there with my hands at some point, making those gestures. It felt intuitive to me, I suppose because we use our bodies to make sense of complicated science—we have forever and ever—and our bodies are the one thing everyone's going to have on them. You don't want to have to send the reader to a video, or to a map or an illustration. Making it literally embodied seemed like the right move.

One moment when I both felt really charmed and also thought, "Oh wow, I'm totally surprised by what this story has done," was when a friend of mine sent me an email saying, "I've just got to tell you, I was just sitting on the subway watching the person across from me do something with their hands that made me realize what they were reading on their Kindle." Which, you know, was probably baffling all the other subway riders.

That's even better than catching someone reading your story in a magazine—catching them actually enacting what you wrote.

Yeah, we all have that fantasy that the person on the plane next to you is going to be reading your story. Which never comes true for me, by the way, which is why I think this was particularly fun and special.

So what was the most challenging part of the reporting and writing?

The reporting really was a pleasure. I just would like to say for the record that I love scientists. They are the best sources ever, and they are so unbelievably generous. A lot of other people too were supremely patient with me, and with making sure that I really did understand this. The emergency people were a real pleasure. This was one of the rare stories—again, a gift to a journalist—where boy, the sources really want you to tell it. You're not fighting a politician who is trying to put his own spin on a scandal. You're not filing FOIA requests. Everybody involved is as invested or more invested than you are in getting the story out.

So that was all great. Writing is hard. It was difficult, as it often is, to decide what not to say. I have volumes and volumes and volumes of reporting on this stuff, and in that way that natural disasters are interesting, almost all of it is interesting. There is a lot to be said about every component of this story. I mean, the amount of information I couldn't put in there is immense, and it was hard to let it go. That was another great gift of having to ignore it for many months. I came back to the story when it was a lot easier to let some of that information go.

Probably the hardest thing about writing this piece was that from the beginning, this story was two stories for me. It's the overt, obvious story, which is the story of the Cascadia subduction zone. On its own, that's an incredible story, one of the best I've ever happened to chance upon. But, from the get-go, in my mind, it was also really a parable about climate change. And then one level deeper than that, it's not a parable, but an example of a really deep problem in our human existence, this kind of problem of scale. We are bound by certain temporal and geographic coordinates, and it's very very hard to see beyond them.

So for me, one of the biggest challenges was figuring out how to weigh those parts: How much to tell it straight as the story of this impending earthquake, and how much to tip my hand about my own feelings about what this story means and what it stands for. And ultimately, I decided to hold my fire. I realized that I could say a lot with just two lines near the end, and that's what I did.

I want to ask you about the reaction to the story. Did you or your editors in any way foresee the scale of the reaction it provoked?

Could you actually please call my editors and ask them that? Because I have no idea if they foresaw it. I definitely didn't. I mean, the piece had been hanging over my head for over a year, so my own feeling about it was like, "Oh thank God, it's finally out of my life." I was not thinking about the reaction. And then right when the story hit, I vanished to central Oregon. A day or two later, when I dragged myself to a little spot of cell service, I got the shock of my life.

So, having watched some of the response unfold, my mental picture of you is that you're standing in the middle of it just like Chris Goldfinger during the Japanese earthquake, looking at your watch while the shaking goes on and on, and thinking, "Holy shit!" So what was it really like to get that kind of reaction to a story?

This place I go to in central Oregon, this little Forest Service cabin in the middle of nowhere, is very free of amenities and outside distractions. You have to walk half a mile to borrow the Wi-Fi signal at the general store. So when I did that for the first time after the story came out, when I first opened my laptop, I think I just stared at the screen for about a minute and a half—then closed it and walked home. I put my head in the sand for a couple days, and then I got my act together and got better cell service.

In addition to not expecting a big reaction at all, I did not remotely expect the nature of the reaction, which was so consistent—people just kept saying the story was terrifying. I wish I could tell you that I've got a future as a Hollywood screenwriter, and that I was making these incredibly deliberate decisions to terrify everyone, but no, not at all! I was thinking about policy and psychology, you know, the philosophical conundrum of being human. I was being nerdy. And while I'm human—I have an ego, and I was happy people were reading the story and talking about it—I felt that the terror and panic were not constructive. It was not the outcome I wanted from a story like this. Fear can be useful, but mostly it isn't.

So initially I was quite worried that there would be this brief reaction—

"Whoa, scary disaster-porn story"—and that would be that. As time went on it became clear that actually, it wasn't being taken as disaster porn, it was being treated seriously. I was really happy that the seismological community and the emergency management community stood behind it. And it seems like people are actually doing things. Contractors are emailing me to say, "You just made my year financially." People are bolting their houses to their foundations. I've gotten enough indications that, in various ways, communities in the Northwest have taken it seriously.

And so, honestly, my feeling now is that—I'm thrilled. It's really rare, as a journalist, to ever feel your work matters in any traceable way. A lot of what you do is on faith, that somehow it matters. Or it matters to you, and that's a cushion. And here's one case where it's like well, maybe it actually matters. I'm so happy to have written something like that once in my life.

PART 3

How Do You Report a Science Story?

YOU'VE LANDED AN assignment, and it's time for the fun part: You get to learn all about whatever captured your interest.

But who do you talk to? Some sources are obvious: If you're writing about a scientific finding, you should talk to the researchers who did the study, and you should interview at least one outside scientist for perspective. And you don't need just any old outside scientist: You need someone who really understands the research, has the perspective and distance to judge its value and relevance—and crucially, who will respond in time for your deadline.

And depending on the type of story you've been assigned, you may need a few more—or many more—voices. The possibilities go on and on: You may need additional scientists with particular expertise, or government officials, or people affected by the medical condition or environmental issue you're covering, or people with traditional ecological knowledge. Your story will only be as complete as the material you gather, so it's essential to reflect on the full range of expertise and experience that may be relevant.

As you look for sources, you need to be careful: If you're not paying attention, it's easy to end up with sources who only represent privileged backgrounds and perspectives. Mostly white, mostly male, mostly highly empowered voices tend to be the easiest to find. And if we allow ease to determine our sourcing, we end up perpetuating the elevation of the privileged, while giving a distorted view of the full reality of the world.

The following chapters explain how to find diverse, knowledgeable, thoughtful sources, and further resources are available at *The Open Notebook*, centering diversity, equity, and inclusion. Finding diverse sources is critical. Handling issues around diversity well requires ongoing awareness and continuing learning, but it's worth it because, ultimately, inclusive sourcing is not only the right thing to do; it also strengthens stories, helping them to better serve readers.

Now that you've found your sources, what do you ask them? To start, you need to come in prepared, both having familiarized yourself with background knowledge and having prepared questions as a guide. You also need to keep the purpose of your conversation in the front of your mind: This isn't just a fun chat; you're here to get the information your story needs. At the same time, a good interview is a conversation, so you need to have sensitivity to the flow and connection and the flexibility to ask good follow-up questions.

Interviews won't be your only source of information. You'll also need to make sense of scientific studies, extracting the information that will serve your story. You may need to go into the field to see for yourself, or dive into data or read government reports.

Throughout the reporting, follow your curiosity—and your skepticism. Your curiosity will lead you to the gems that will capture your readers' fascination. And your skepticism will help you check and recheck the conclusions you're drawn to, building the storehouse of evidence to ground your piece in facts.

Siri Carpenter

17 Is Anyone Out There? Sourcing News Stories

Geoffrey Giller

IF YOU'VE EVER written a news story on a tight deadline, you've probably had some version of this thought: Please, will someone—anyone—call or email me back?

The gap between getting a news assignment and pinning down sources can feel daunting. You have little time to seek out suitable sources, and once you find them, the pressure is on to make the most of the interviews. But there are ways to make the news-reporting experience less stressful.

Finding Sources

You've got the assignment—let's say, a 700-word news piece, due tomorrow, about a new study in a scientific journal—and you've done your background research, including reading the study and related material. Now it's time to find credible sources, and fast. For a news story, you want at least two, preferably three (sometimes more). If you're reporting on a new study, your first source is pretty much set: either the first author or the corresponding author.

Identifying a good second source can be trickier. Ideally, it's someone who's not directly involved in the research but is familiar with it—or at least with the field—and who can offer a different perspective. The primary source can help point you in the right direction. But as Andrew Grant, a physics writer at *Science News*, cautions, "You have to watch out, because they're more likely to suggest somebody who would think that their work is awesome." To avoid this trap, Michael Lemonick, a writer at *Climate Central* and former senior writer at *Time*, sometimes asks a study author for the name of someone whom the author respects but who might not agree with her or him.

A study's references section can point to good secondary sources, as can other past news coverage of the topic. Google Scholar can also turn up review papers in the relevant area, whose authors might make good sources. Eventually, as you become more familiar with a particular scientific discipline, "you get an idea of experts in different sub-areas of your beat" and develop a "Rolodex" of sorts, says Grant.

It's usually best to contact sources by email to schedule interviews, but as Grant notes, there are circumstances when he phones instead— for instance, if his deadline is extra tight or if he's writing about a major paper that's under embargo until its publication, in which case the study's authors probably expect a lot of media calls.

To secure a second source, sending out multiple requests at once can keep you from having to wait on one person who may or may not respond. If they all respond, that's more material to work with. "If I have to talk to more people than I need, that's not bad," says Lemonick. Some sources provide more helpful comments than others, says Becky Oskin, a senior writer at *LiveScience*. Plus, says Grant, extra interviews give you a chance to make new connections and sniff out other story ideas.

Casting a wide net is also useful for those times when you discover that a source has an agenda. If someone is overly negative about a study or its author—or a little too positive—then you'll be glad to have an additional interview lined up.

Getting a Response

Crafting a subject line that will get a busy scientist to open your email right away, rather than after your deadline, is as important as what you put in the email itself. If you're like Lemonick and you write for widely known publications like *National Geographic* online and the *New Yorker*'s website, sticking the publication name in the subject line gives you an advantage. "People who are inundated with requests for interviews might well prioritize publications they think are well-known," he says. Otherwise, writing "media request" or "journalist request" is a good way to make sure a researcher knows the inquiry is from a reporter who needs a quick response, says Oskin.

The email itself should be short and sweet, adds Oskin. "I tell them who I am, I tell them what I want, which is usually a phone interview, and I tell them when I want it. And that's it—get in and get out." She also finds that if she mentions that she's read the paper in question, scientists are more likely to respond.

But even if you're writing for a marquee publication and your email is direct and to the point, some sources may not respond as quickly as you'd like. When that happens, Lemonick sends a follow-up email, sometimes including a few questions in case the source can quickly fire off some answers. Remember also that you're not limited to email and phone: Oskin says she once messaged someone via Twitter when he didn't respond to emails or answer his phone.

On the Phone

Once you do get a source on the phone, you may not get a second chance to speak with him or her, so it's important to make the most of the conversation. If you're covering a new study, make sure you've read the whole thing before the interview so you can clarify anything you don't understand. And having a list of specific questions or points to go over before the call can help ensure you don't miss anything crucial. For many pieces, Oskin uses a standard set of questions that she can tweak and tailor to the particular study or topic: How did you get involved in this project? How might I sum up these results in just one or two sentences? Can you provide some context for me to understand these findings?

When you're talking with a source, pay close attention, listening particularly for eloquent explanations, appealing turns of phrase, or arresting ideas. "You can hear a good quote go by," Oskin says. Be sure to get those gems down right away.

With a tight deadline looming, it might be tempting to keep the conversation as brief as possible. But Oskin notes that cutting a source off too early can mean missing a nice quote that sums up their thinking. "Usually, people say the most interesting stuff when they're finishing up their thoughts," she says. "They've finally organized themselves into getting out the most pithy sentence, or the thing that actually summarizes everything that they've just said." At the same time, Lemonick says that if a source has already given you good, interesting quotes, "the idea

Go-To Interview Questions for News Stories

- How did you get involved in this project?
- What did you find?
- Why is it important?
- Did you encounter any surprises along the way?
- Were there any unexpected hurdles?
- What are the next steps for this research?
- Are there any specific questions or criticisms others might have about these findings?
- How might I sum up these results in just one or two sentences?
- Is there anything I haven't asked that you would like to add?
- What do you personally find most exciting or important about these results?
- Can you provide some context to help me understand these findings? How do they fit in with other recent results?
- For field studies: Describe what you saw during the fieldwork.

(*Sources*: Becky Oskin and *The Open Notebook*. Download a print-friendly version of this list at bit.ly/TONnewsQs.)

that they might say yet one more interesting thing is not so crucial." He typically wraps up the interview when he has what he needs, with the requisite final question of any interview: "Any last things to add?"

From Notes to Quotes

After an interview, Grant immediately goes through his notes to mark the "money quotes" and important explanations. He'll also flesh out any thoughts he had or notes he made during the call. Both Grant and Oskin say they rarely record calls for news articles, instead relying on their notes.

Oskin says she's gotten her science writing down to a science. "It's a lot of work," she admits, but her previous experience working at a daily newspaper helps her turn around well-written pieces at a blistering pace—she writes up to two news stories per day. Grant, who moved from the monthly *Discover* to the daily, online *Science News* two years ago, says he feels the challenge of being among the first to report a topic and make sure he covers the whole story. But, he says, "It is really rewarding when somebody highlights it or you read *Discover* or something and see that they built off your writing."

Of course, after the interviews are done, you have to actually write the piece. But that's a discussion for another day.

18 Interviewing for Career-Spanning Profiles

Alla Katsnelson

PROFILES OF SCIENTISTS in the golden years of their careers take the measure of a life in science and reveal the motivations that have guided the scientist's work. Conducting interviews for such career-spanning profiles in a way that elicits both the telling little details and the big themes of a person's life can be tricky. For one thing, a lifetime is a lot of ground to cover. And it can be hard to find the details that truly shed light on a person, and to do so without being unfairly intrusive or resorting to cliché.

Yet these stories can also be deeply rewarding, precisely because the interviewing is often so intimate. When done well, "legacy" profiles reveal something that's usually hidden: how the swirl of a person's inner world connects with the accomplishments they make in their outer world.

"A legacy story in my mind is basically an advance obit," says Jacqui Banaszynski, a Pulitzer Prize–winning journalist who is a professor emerita at the University of Missouri School of Journalism and a faculty fellow at the Poynter Institute. "You're gathering information when the person still has a chance to reflect on their life."

The secret to conducting good interviews for career-spanning profiles lies in the questions you ask and the connection you build with your subjects. "We have to turn our story subjects into storytellers—most naturally are not," Banaszynski says.

Where to Begin? (Often, the Beginning)

A successful profile weaves together three parallel timelines that make up a subject's life, says Banaszynski.

The first is the subject's basic biography and "résumé stuff," including—in the case of scientists—basic information about the subject's most noteworthy scientific accomplishments. The second timeline includes defining personal moments in the person's life. And the third is the social and historical context of their work. (For a recent example of such interweaving, see Natalie Wolchover's 2015 profile of physicist Nima Arkani-Hamed, published in *Quanta* and featured in *TON*'s Storygram annotation series.) These strands require different types of reporting—some of which can be done before the interview.

Come interview time, you should already know as much as possible about the first timeline—where the scientist has lived and worked, what their big discoveries were, when they published their key papers. You might even want to write this timeline out and bring it along to your interview, Banaszynski suggests, to use as a jumping-off point into the other two.

It's easy to let the interview get bogged down in the basics of the subject's biography or technical details of the science. That material is already accessible in the literature, though, so try not to spend much valuable interview time on it. Instead, keep your focus on the second and third timelines—for example, how or why specific studies became a springboard for the subject's overall research program, or how starting their career at a particular time or place contributed to their interests.

A scientist's "origin story"—the beginning of the second timeline—is one obvious starting point for a profile interview, says Robin Marantz Henig, a freelance science writer and a contributing writer for the *New York Times Magazine*. She spends a lot of time on questions such as *What were you like as a kid?* or *When did you first realize you were interested in science?*, because people generally don't talk much about their childhood and early history, so the response to such questions is likely to be relatively unrehearsed. Also, unless you're speaking to a celebrity scientist, such personal background is likely not to have been reported before, so it can yield some rich material.

But "origin stories" aren't always the best approach to starting a profile interview. For one thing, not everyone feels comfortable opening up about personal experiences. "Friends of mine have had people say, 'If you're going to ask me personal questions, I'm just not going to talk to you,'" Henig says. "I haven't had that, but I've had people who just haven't had that much interesting to say about their youth." In such cases, starting the interview by talking about their research or their early-career interests or accomplishments can loosen up the flow of conversation and help build rapport—and doing so can eventually open doors to more personal introspection.

Go Deep

Whatever your takeoff point, there are some standard questions that are essential to any career-spanning profile interview. *How did you get your start? How do you do your work? What is your typical day as a scientist like? What inspired you to pursue [whatever questions] in your work?*

But running through such questions in a rote way can yield stale material, cautions Yudhijit Bhattacharjee, a contributing writer at *National Geographic* who also writes for the *New Yorker* and the *New York Times Magazine*. "Scientists generally have pat responses to these questions—they know you're going to ask them, and they've probably said similar things to other journalists," he says. "Anything you're getting that sounds clichéd is probably only half true."

When he hears anything in these responses that's unexpected, he'll drop the script and drill down. "You've got to read between the lines," Bhattacharjee says. "When a scientist says something like, 'Well, back then this method didn't work,'" he says, "you can't gloss over that. Something there might be revealing if you investigate." Probing further (*What do you mean it didn't work? Why did it work later?*) can take the conversation to a deeper place—say, to the topic of perseverance, or of a changing cultural context that affected how the science could be done—and perhaps even open up a discussion about the subject's personal character and how it ties to their science. "They have to see that you are after something that is not superficial," he says.

Bhattacharjee's strategy dovetails with one of Banaszynski's favorite rules of interviewing: For every answer you get, ask five more questions. "The first answer will probably be very general," she says. "Stay in the moment and peel it back." Every answer can be plumbed to go a step further, she says, if the reporter is listening for promising nuggets and willing to linger on them.

Key to the success of this strategy, Banaszynski adds, is allowing enough time for those nuggets to emerge. If the person you're profiling doesn't reply to a question right away, don't rush to the next one. Force yourself to stay silent to provide the space for them to answer, even if it feels awkward; scribble in your notebook if you have to.

Ask about Turning Points, Failures, and Oddball Details

A profile is not just a résumé in paragraph form. What you're probing for in a legacy profile are the personally meaningful turning points in the subject's career. "What were the moments that led them to this life, and what was it about those moments that made them special?" says Banaszynski.

As a newspaper reporter and editor, she says, for her those moments might include the first time she saw her byline in print, say, or the first time she got a big scoop; ask about "the science version of that." Some subjects will need more massaging than others to understand what you're after, so if you don't get a satisfying answer, she says, ask a slightly different question that might help them reply in a more interesting way. One variant here might be: *What isn't in your résumé that you think should be, in terms of its importance to you?*

Another place where turning points often lurk is in failures, or "moments of stuckness," Banaszynski calls them. You might say: *What were some moments in your career that were especially challenging or discouraging? Tell me about a time when it didn't seem like your project would work? When have you felt like quitting?*

While you're seeking out these big-picture themes, don't ignore the close-up view, says Bhattacharjee. "Sometimes I'll ask these stupid little detail questions, just trying to see if it might lead to an interesting avenue," he says. For example: *Did you get the phone call when you were sitting in the office, or were you somewhere else?* "Sometimes they'll say, 'Oh yeah, actually, funny story . . .'" And who knows where that story will take you?

Don't forget to look around you for details that can liven up an interview, too. For a profile in *HHMI Bulletin* a few years ago, Henig asked the researcher she was interviewing about a plant in his office that was wearing a pair of sunglasses, and learned that the plant was a gift from one colleague who thought its leaves looked like the mop of hair on another colleague. Henig had been struggling to move the interview into evocative territory, and that detail offered a fun and revelatory snapshot of lab dynamics. "[The researcher] was looking at me askance, like, 'Why are you asking me so many questions about that? I'm telling you about my brilliant science!'"

Indeed, says Banaszynski, profile subjects often don't understand what material journalists need. Letting them into the process can help them feel more at ease and in good hands. It may help to explain before or during the interview that some of your questions might seem weird or overly specific, but that the aim is to help you paint a picture in your head so you can help readers to really envision what the subject experienced.

Henig describes having to return to a profile subject because her editor told her she hadn't brought the scientist to life vividly enough. "I actually told the subject that, which made her a little uncomfortable but turned it into a problem we needed to try to solve together," Henig says. She explained that she would have to ask more questions to capture the scientist's response to a particular pivotal moment in her life. Eventually, key details emerged.

Regardless of how you decide to structure the interview or what you decide to ask, do be an active participant. Another of Banaszynski's rules of interviewing: "Don't be boring." Laugh when it's funny; show your surprise. "Anything you can do to show you're listening," she says. In order to move past the subject's well-rehearsed answers—the predigested story of our lives that all of us have at the ready—you need to spark their interest by engaging with what they are telling you, not just transcribing it at face value.

Dull Interview? Don't Be Afraid to Shake Things Up

If an interview with a profile subject is going well and you're getting the material you need, there's no reason to interfere with the flow. But if an interview seems to be falling flat and you know you're not getting what you need, you might have to shake things up.

There's no single way to do so. Bhattacharjee says for him, challenging the subject somehow—even to the point of seeming aggressive—can turn the tide. Or, he'll frame a bit of pushback as a joke: "How could you have gotten these results in a year? Did you have a bunch of graduate students crunching the numbers while you were in Cancun?"

It sounds provocative, he says, but the idea is to convey that you have a reason to ask what you're asking. Scientists sometimes don't have a great impression of journalists, he notes. "They don't think you're bringing very much to the table—it's just that they don't have time to write their own story." It's up to reporters, Bhattacharjee says, to demonstrate that we want "to take a deeper look at their humanity. Once people get it, once they realize that you actually do want that, then something changes," he says. "They shift gears and say, 'OK, I thought this would just be a publicity op, but you actually do want to understand me [and] understand the process of science.'"

A less confrontational way to change the tenor of a lackluster interview is to ask the subject to respond to what critics have said—or might say—about their work, Henig says. It can also be helpful to change the subject—maybe by throwing in a more impersonal or less intense question or two. You might hit more fertile ground, and the break can also allow earlier topics to percolate, making them ripe for revisiting later.

If the interview does end up rather flat, remember that you can fill in a lot through interviews with secondary sources—the scientist's colleagues, mentors, students, or family members. Be direct in asking them for what you need, says Banaszynski. "You can even say, 'He didn't want to get personal, so I need some help here.'"

But secondary sources sometimes need even more guidance on what type of input a reporter is looking for, and prompting them for good material can be its own art. If you simply call at the appointed time and ask, *Can you tell me about so-and-so?*, you might not get much, Banaszynski cautions. A day or two before you've arranged to speak, get the source's memory bubbling by emailing them some examples of the kinds of questions you might ask.

A Sense of Place . . . and Time

A major determinant of how your interview will go is where it will occur. "If there is any way to do the interview in person, do that—it makes such a difference," says Henig. "You ask questions differently. The whole pace is different."

After the sit-down part, try to get your subject on the move, perhaps by taking a tour of their lab or another place that's meaningful for them. "If it's at all appropriate, one great place to get people talking is in the car," says Banaszynski.

If you're relegated to the phone, ask setting and sensory questions, says Banaszynski. *What does your lab look like? Are there smells to these chemicals? What's the first thing you see when you step out of the tent in the morning?* Also, ask to see artifacts, or photos of them: the tiny gold grill that holds samples in electron microscopy, the moldy petri dish, the DIY tool used to conduct the surgery.

Finally, if at all possible, try to meet a second time. "One advantage is that by then, I might actually have an inkling of what the article will look like—whether it will be a more personal or a more intellectual kind of thing," says Henig. Coming in for another chat allows her to home in on the details she needs. In addition to giving a reporter a second chance to gather more information, Banaszynski says, the break gives the researcher a chance to mull over the first conversation—and in the meantime, they might think of things they wish they'd said the first time. They're also likely to let their guard down a bit for a second conversation.

If setting up a second meeting (or even a phone conversation) isn't possible, you can create a bit of the same effect toward the end of your first interview, using a trick Banaszynski calls the Columbo technique. Start to wind down the interview, maybe even say your thank-yous and goodbyes. Then take a breath and say, "I know you have to go, let me just look through my notes. . . . Oh yeah, remember when you said X? Can you tell me a little more about that?"

Bhattacharjee, too, stresses the importance of time—"not the duration of an interview, but the interactions over a certain period of time really add dimension," he says. If you actually have a few days on site

A Cheat-Sheet for Career-Spanning Profile Interviews

Below is a crowdsourced list of questions that can work well during interviews for career-spanning, or "legacy," profiles. Of course, it's unlikely that
you'd have the opportunity to ask all or even most of these questions in a
single interview, but they can serve as a framework as you come up with the
ones most relevant for your subject. (Download a print-friendly version of
this guide, which fits inside a reporter's notebook, at bit.ly/TONprofileintvw.)

Early Years

- What were you like as a kid?
- When did you first realize you were interested in science?
- How did you get started in science? In this field?
- How did you get interested in this specific topic?

Turning Points

- How did you discover the problems that led to your most important work?
- What were some challenging or discouraging moments in your career?
- What have been some moments of struggle, uncertainty, or failure?
- What are some things you did that you didn't think would work, but
  did—or that you thought would work, but didn't?
- What were the big surprises along your career path?
- What do you wish you had done differently?

The Work

- What is your group working on at the moment?
- What is your team like?
- What is a typical day like when you're working on X?
- What aspects of your research are you most proud of?
- Who are your mentors or people who have otherwise influenced your
  work?
- If you could alter one aspect of your field, what would it be?

Setting

- What does your lab or field site look like, smell like, and/or sound like?
- What's hanging on the walls of your office?
- What do the tools you use look like?

Life Outside Work

- What would you have done for your career if not this, and why?
- If you could have a completely different career for a week, what would
  it be?

(continued)

A Cheat-Sheet for Career-Spanning Profile Interviews (*continued*)

- Have you ever brought your family to the lab?
- Do you talk to your children or grandchildren about your work?
- When you go out with scientist friends, what do you talk about?
- What are some surprising things that people might not know about you?

Context

- What are some of the main changes you've seen in your field during your career?
- How does the academic culture differ between the institution where you work now and where you worked before? (And for scientists who work in a different country from where they were born: How does the academic culture in the country where you work differ from the one in the country of your birth?)
- What are you looking forward to seeing in your field in the next 10 years?
- What are you looking forward to that you couldn't have predicted a decade ago?
- What big mysteries in your field remain?

with your subject, create opportunities to cross paths. If you planned to meet for a few hours every morning for three days, for example, find a reason to pop by for a moment in the evening, perhaps to return something you've borrowed. "It's almost like a scientific exposure type of equation," he says. "With every exposure you're just increasing your chances of getting better material."

19 How to Conduct Difficult Interviews

Mallory Pickett

IN THE LATE 1990s, Martha Mendoza was a relatively new reporter at the Associated Press. She had just discovered what she hoped would be one of her first big scoops: She had documents proving that Bureau of Land Management field officers had been rounding up wild horses and burros on BLM land and sending them for slaughter, while recording them as adopted.

So, relishing the role of gumshoe reporter, Mendoza set out to do some hard-hitting journalism. She went to the source: She found the corral where the animals were kept. The officers weren't there, so she went back to their headquarters. The front door was locked but the side door was open, so she let herself in. Inside, she found all the officials she had hoped to interview, looking surprised.

"I was like, 'Hi, oh I'm so glad you're all here together. I've been looking for you,'" Mendoza recalls. She dove right in, telling them about the documents she had and what they proved.

"They were like, 'No, get the hell outta here. That didn't happen, you're a liar,'" Mendoza says, laughing. "It just kind of was a gotcha moment that didn't really tease out very much."

Twenty-one years later, Mendoza, who's still an investigative reporter for the AP, has won two Pulitzer Prizes—one for her participation in investigations into a secret American massacre of civilians in the Korean War, and another for a story about slavery in the seafood trade. If she were to try that BLM interview again today, she says, there's a lot she would do differently.

For starters, "I would have role played the whole thing," acting out her questions and the various answers they might elicit. That's something she and her team at the AP do now before difficult interviews. Also, she says, "I would have taken a long time letting them tell me about the program, asking lots of questions," to get the officials comfortable and show that she'd done her research.

These are just two of the strategies Mendoza has developed over decades of investigative reporting and thousands of difficult interviews. But a typical science journalist doesn't always get this kind of practice. When you're reporting on science, interviews are often friendly—closer to collaborative than combative. Interviews that are more adversarial can be intimidating, and, in the worst case, they can be a deterrent to taking on investigative science stories. While the best way to get over this fear is probably to amass many years of practice, like Mendoza has, science writers with less investigative experience can fast-track their progress by learning from the pros.

It's OK to Be Nervous, but It's Essential to Be Prepared

Every journalist I interviewed for this story, even the most seasoned investigative reporters, said they dislike confrontation and get nervous or uncomfortable before what's likely to be an adversarial interview. To make sure nerves don't interfere with your work, it's essential to be prepared.

Stephanie Lee covers science for *BuzzFeed News*, and over the past few years she's broken several stories that uncovered various kinds of malfeasance in science: sexual harassment, harmful drugs, and dubious research results. She is frequently on the phone with people, trying to get them to talk to her about their worst misdeeds and mistakes. In 2017, she wrote a story about a biotech start-up founder who had been representing himself as having a PhD from the University of Pennsylvania, when in fact he didn't.

One way that Lee combats the nerves that inevitably come upon her when she's about to ask a hard question is by writing it out verbatim—so that when she asks, for example, "Do you in fact have a PhD from the University of Pennsylvania?" she just reads it from the screen in front of her.

"In the moment, it can feel very paralyzing to formulate and ask a very accusatory question," she says.

Robert Cribb, an investigative reporter at the *Toronto Star*, uses a tactic similar to Mendoza's interview role-play. Before sitting down with an important source, he'll create a flowchart, starting with his first question and mapping out every possible answer, and his follow-up questions. Then he sits down with his editor and they'll act out different routes through the chart.

It's a lot of work, but Cribb says that after he's done this, he's supremely well prepared not just for the interview but also for writing the story.

Mendoza sometimes brings what she calls a "dossier" with her to an interview—a compilation of relevant documents and information that she's gathered, which she gives to the interviewee at the beginning of the interview to help them process what's happening, and to show that she's done her homework. (This has the added benefit of buying a little more time with someone who might be trying to close the door on her if she's showed up for an impromptu interview—she gets at least a few more minutes with them as they review the dossier.)

Interviews with the main subject of an investigative story shouldn't be about fact-finding, she says. Of course, you should keep an open mind to any new information that's provided, but in general it's essential to really know your stuff before you start. The purpose of the interview is to get the subject to respond to the facts that you've gathered, and to hear their perspective, the reasoning behind their actions, or perhaps their flat-out denial.

Staying rooted in your reporting can also help prevent a conversation from getting too heated.

"You have one thing as a journalist in these conversations—you have the truth of what you have gathered," says Cribb.

Mendoza says that when things do get heated or sources start to shut down, she always tries to give them a chance to explain themselves. This strategy is also useful for dealing with sources who turn suspicious mid-interview, or get nervous, after the fact, about what they've said. In all of these cases, she says the best thing you can do is to ask interviewees to share their concerns by saying things like "I sense that you're uncomfortable talking about this topic. Would you be willing to share why you're concerned?," "Do you want to know what the story is going to be about?," or "Let's talk about what you're worried about."

Fight Hard to Get All Sides of a Story

Sometimes the most difficult part of interviewing the subject of an investigative story is just getting the interview at all. Cribb says he always starts with a friendly introductory email, and follows up with phone calls if he gets no response.

Every time he reaches out, he reminds the subject that it is in their interest to get their side on the record, and that he is doing his best to be fair and accurate. He says this works for him "about 47 percent of the time," but he's noticed it getting harder over the years. "There's far more manufactured, manicured messaging in this day and age. The public relations industry is far bigger and more sophisticated than the journalism industry," he says.

But it's important to keep trying. Lisa Song is a Pulitzer Prize–winning investigative environmental reporter who recently joined *ProPublica* after several years at *InsideClimate News*. In 2014, when she was at *InsideClimate*, Song and a team of her colleagues were working on a story about a weakening of air-pollution guidelines in Texas that essentially gave oil refineries and other polluters license to emit more benzene. The team worked on it for over a year, periodically reaching out to environmental regulators requesting an interview. They consistently refused, and would only interact over email.

Finally, after a year of reporting, Song sent out a routine email asking again to speak with someone, not expecting much. But suddenly one of the toxicologists from the Texas Commission on Environmental Quality agreed to meet with her in person. She rushed onto a plane to Austin and was rewarded with a one-hour sit-down with him and several of his colleagues.

"I [laid] out all of the concerns from other scientists I talked to, all of the data and exposure limits from other agencies in the country, all of the science, to say, 'This is what everybody else says. This is what the current state of science is saying. What do you have to say?'" Song recalls. The official's response was to simply state that he believed the limits in Texas were protective enough.

"It was a difficult conversation in the sense that he refused to recognize the facts," Song says. But, she adds, "I think it showed the depth of his conviction." He was an influential figure in the story, and by meeting him in person she was also able to add some details and nuance to the piece that wouldn't have come through over the phone.

When you don't have the luxury of waiting a year for a source to come around, your best persuasive efforts may not be enough to get the interview you want. Instead, subjects may only be willing to send a written statement in response to a list of your questions. Formulating these questions is just as important as preparing for an interview, so there are a few things to keep in mind. Cribb says to remember that just as their responses are on the record, so are your questions. So be mindful of how you phrase them, trying to stay as neutral as possible. Be as specific as possible, and make sure you haven't put two questions in one question. "Don't say, 'Is this true? And if not, what about this?'," Song says. "Break down your questions into simpler pieces so it's harder for them to try and miss a question."

Additionally, it's a good idea to include a note with your questions letting the source know that you're disappointed they wouldn't agree to an interview, and that this fact will be noted in the story.

Most importantly, remember that reaching out to the "other side" isn't merely a formality. Getting everyone's perspective is an essential part of the reporting. So whatever your sources say, keep an open mind.

Mendoza says talking to the subject of an investigation "always changes the story. Usually, I think I've got the story and then it's like, oh, I've got to do some more," Mendoza says. "It's not 'gotcha.' Everything is nuance."

No Surprises

Lee says that during the reporting process, even her most persistent emails and phone calls often get no response. But before publishing any investigation, *BuzzFeed News* always sends out a "no surprises" letter to the people or institutions in the story. This is a list of all the facts and new revelations uncovered in the story. (See a sample at bit.ly/TONno surprise.)

Song and Cribb have similar protocols. Cribb says that at the *Toronto Star*, they won't include any allegations in the story unless they've run

A Cheat-Sheet for Difficult Interviews

- Don't fear the fear! It's natural to be nervous before an important or difficult interview.
- Prepare. Make sure you have all your facts and reporting in order and that they will be accessible to you during the interview if you need them. Then practice the interview as much as you can—map out an interview flowchart and/or role play with a colleague.
- Stay calm and stick to the facts. During the interview, if your interviewee starts to get upset or shut down, ask them to share their concerns. "I can tell that you're upset/reluctant to talk about X. What specifically are you worried about?"
- Make every effort to include all voices in your story. Don't just send one email and then say the source didn't respond to a request for comment. If you can't get a phone or in-person interview, put serious thought into writing questions for written response. Remember that your questions are on the record; keep them neutral.
- No surprises! If you work with a publication that has a fact-checking department, they will run much of your reporting by the source. If you don't have a fact-checking department, a "no surprises" letter that goes through everything you said about that source and gives them a chance to respond can be useful. A "no surprises" phone call can also be a great idea, but make sure you record it.
- Be kind and fair. The relationships you build with the people you're investigating will make reporting easier, and could even lead to new stories down the road.

them by the subject of reporting. So if he realizes he has six points to make and he only asked the subject about five of them, he will go back and talk to them again, or, if there's no time, he won't include that sixth point.

In addition to being a necessary courtesy to the people you're reporting on, an added benefit of a "no surprises" letter or phone call is that it can sometimes lead to unexpected last-minute interviews. "I've found that by being upfront and giving these types of opportunities, that sometimes you can get way more conversation than you expected," Song says.

Everyone appreciates fair treatment, and being open, honest, kind, and fair can often help avoid animosity altogether. Even if the subject of your investigation hates your story, you may find they respect you and your process. Cribb has even had previous subjects of hard-hitting investigations come to him years later with tips for new stories and offers to help.

Don't Let Fear Keep You from Telling Important Stories

Lee says it's taken her a long time to get used to investigative work. "These kinds of stories are not natural for me. It's really taken a lot of practice," she says.

She admits it can still be scary for her to take on investigative, potentially confrontational stories, but "I do them in the hopes that people will have a better understanding of what's really happening," she says, "and I try to hold companies or scientists accountable if they're not telling the truth."

Song agrees, and she hopes more science reporters will feel empowered to tackle awkward interviews and take on investigative work.

"Science journalism is a hard thing because there's an extra layer of stuff to decode," she says. "I think it's totally ripe for investigations into abuse and twisting of truth and facts, and how they influence policy. So the more [investigative] stories that are out there, the better."

20 Including Diverse Voices in Science Stories

Christina Selby

A TEACHER STANDS in front of a classroom and instructs five- to seven-year-olds to take up their crayons and draw a surgeon, a fighter pilot, and a firefighter. When a surgeon, a fighter pilot, and a firefighter—all women—later walk into the classroom, the children look stunned. They had drawn 61 pictures of men, and only five of women. A video of the classroom exercise, produced by a British charity, calls attention to the early age at which gender stereotypes take root in children's minds.

Science writers use words to draw pictures every day: pictures of biologists, physicists, technology developers, physician-researchers. Our readers' mental images—and our own—may not differ much from those of schoolchildren. According to the Global Media Monitoring Project (GMMP), a nonprofit effort to track women's representation in media stories, women are the central focus in only 14 percent of science and health stories, and only 19 percent of experts quoted are women. Stories that challenge gender stereotypes—by overturning common assumptions or representing women in counterstereotypical roles or situations—made up just 5 percent of science stories in 2015.

People of color are likewise poorly represented in the news media, including in science stories. "Whenever communities [of color] interact with science in the news, it's rarely positive," says Danielle Lee, a biologist and *Scientific American* blogger who writes and consults about diversity in the sciences. "You have to be super young or do something amazing for there to be a positive story." That's true across the spectrum of science, technology, engineering, and mathematics (STEM), and it creates a narrative, Lee says, that "our participation is rare or special"—when in fact, "there's plenty of us out there."

Media stereotypes shape societal attitudes, behavior, and expectations. "Media tells us our roles in society—it tells us who we are and what we can be," writes Julie Burton, president of the Women's Media Center, in the organization's 2015 report on gender inequality in the media. And the early establishment of stereotypes affects whether girls choose a science profession and whether young women stick with their education to become working scientists, says Nilanjana Dasgupta, a social psychologist at the University of Massachusetts at Amherst who studies implicit bias in STEM professions. Women make up only 29 percent of the scientific workforce, according to National Science Foundation data. In the fields of engineering and physics, that number drops into the single digits. NSF data also show a race gap in STEM fields; Blacks and Hispanics together make up only 11 percent of the STEM workforce.

When science writers draw pictures of science in action, whom do we draw? What is our responsibility to represent the current diversity of scientists and avoid perpetuating stereotypes that, among other things, may contribute to lack of diversity in STEM fields?

For science writers who want to be part of the solution to the lack of diversity in science journalism, here's a four-step program:

1. Recognize (often hidden) biases that may get in the way of including diverse voices in your stories.
2. Track the diversity of your sources.
3. Break out of old reporting patterns to cultivate new sources—and be an ally to underrepresented communities.
4. Broaden your definition of "expert."

Recognize Your Biases

Journalists' beliefs and attitudes play a significant role in determining whose stories are told and whose are left untold. We all hold stereotypes about people based on their gender, race, ethnicity, sexual orientation, religion, and other group memberships.

Some of these biases are so deeply ingrained we don't recognize them; and yet these implicit biases can lead even the most egalitarian-minded among us to make assumptions about who an ideal scientist or expert is. "The image that pops into most people's minds are historic figures—typically men—mostly white or sometimes Asian," says Dasgupta. "These implicit biases," she says, "end up not just being an image in our minds, but also influence our decisions and actions." The action may not be conscious, but if a person doesn't fit our mental image of an ideal scientist, we may not call on that person as a source.

Veteran journalist Howard French recounts the power of implicit biases in newsrooms in his *Guardian* article "The Enduring Whiteness of the American Media." Black journalists, he says, are "heavily concentrated in stereotypical roles"—such as reporting on entertainment, sports, and urban affairs—and often excluded from covering many other areas, including politics, national security, big business, and science and technology. A case in point: In the early 2000s a Black journalist with over 10 years of experience at the *New York Times* successfully pushed to cover the emerging tech boom. "There was immediate pushback from his white colleagues, who claimed he had no background in tech and was not the right person for the job," French writes. "A black man occupying this space did not fit preconceptions."

Not much has changed in newsrooms in the years since, he says. The fact that both the journalism industry and the scientific workforce continue to be predominantly white and male "is not a trivial matter, or a subject of concern solely to journalists," he writes. "The overwhelming whiteness of the media strongly but silently conditions how Americans understand their own country and the rest of the world."

A necessary first step in countering one's own social biases is recognizing and acknowledging that they exist. You can assess your own biases through the tests at Harvard University's Project Implicit. The testing starts with a disclaimer: "If you are unprepared to encounter interpretations [of your test] that you might find objectionable, please do not proceed further."

The good news is that implicit biases are malleable. We can gradually retrain our brains and unlearn biases. That's to say, we can change our image of who we believe fits in science, and therefore whom we call as sources for our stories. The Bias Cleanse, produced by Ohio State University's Kirwan Institute for the Study of Race and Ethnicity, is a good place to start.

Track Your Sources

Countering the lack of diversity in science stories might mean examining and making changes in sourcing. Science writer John Platt was reporting for *Motherboard* on gender inequality in technology when he learned from Kate McCarthy of SheSource, an online database of female experts, that fewer than 20 percent of expert sources cited in political news in major US print media outlets are women. Struck by this statistic, he wondered how his own articles—mostly covering the life sciences—measured up. He resolved to achieve a fifty-fifty split. Since January 2015, he's made a point of using more female sources. Platt was well on his way to his goal when a feature assignment on tech skewed his numbers down again: He was dismayed to discover, too late, that he hadn't interviewed a single woman.

Adrienne LaFrance, who writes about science and technology for the *Atlantic*, also recently shared the results of her second year-long attempt to include more women in her stories. The result, she writes, were "distressing": Her year-two numbers were lower than in year one. She realized her efforts were only getting her so far. "Some people would argue that I'm simply reflecting reality in my work," she says, referring to the low numbers of women who work in tech. "That's an overly generous interpretation," she writes. "By substantially underrepresenting an entire gender, I'm missing out on all kinds of viewpoints, ideas, and experiences that might otherwise sharpen and enhance my reporting."

Learning to consistently include diverse voices is "a career-long process," says Albuquerque-based freelance science writer Laura Paskus, who covers natural-resource and social-justice issues for *High Country News*, *New Mexico in Depth*, and other publications. Issues such as suicide among youth or climate change often disproportionately impact minority communities. Paskus knows that reporting accurately on these issues in a minority-majority state requires including the people most affected—Hispanic and Native American communities, which in New Mexico make up 46 percent and 8 percent of the population, respectively.

"Meeting with the people whose communities are being affected not only deepens the story I'm already working on, it inspires more and more story ideas," says Paskus. Even though she has purposely changed her reporting practices and pays attention to who has a voice in her articles, Paskus—like Platt and LaFrance—is "often disappointed when I look back on a previous year's work and realize that I could have done much better at including diverse voices in many stories."

As Platt, LaFrance, and Paskus have found, individual reporters' attempts to include more diverse sources can be stymied by structural inequity—the policies and practices of institutions in science and technology, and in the field of journalism, that favor one group over others. "Understand that the best way to get diverse sources is to have diversity in your newsrooms," says Roberta Rael, director of Generation Justice, a social-justice nonprofit that trains teens in journalism.

For freelance writers, understanding the influence of inequity at different levels in our society is essential when setting goals to change your reporting practices, says Rael. Women and people of color represent a small percentage of professionals in many STEM fields because of institutional and cultural biases. Working against the tide of structural inequality requires science journalists to make ongoing efforts to find and tell these minorities' stories.

Break Out of Old Reporting Patterns—and Give Back

Cultivating new sources can take time. Turning news stories around on a tight deadline doesn't always lend itself to elaborate new source-finding methods. But that doesn't mean it's impossible to prioritize diversity in sourcing when time is short. It's often possible to improve diversity in sourcing just by remembering to pay attention to your choices. And feature writing offers more lead time, and so even more opportunity to cultivate new sources.

Platt reviews old stories for sources he's used in the past, uses She-Source to find new sources, and regularly attends local nonprofit and community events to increase his network of contacts. He also asks sources for their personal backstories: how they became interested in the topic at hand. This approach, he says, can yield new stories or vastly different perspectives on the same story.

Relying heavily on past media coverage to find expert sources can perpetuate the exclusion of underrepresented groups. Breaking out of that mold requires casting a wider net and searching more systematically for sources. Organizations that serve underrepresented communities can be good conduits to diverse sources for some kinds of stories. Sarah Gustavus, a radio producer in New Mexico, reaches out to professional groups such as the Hispano Chamber of Commerce, Native Public Media, and minority student science groups and their advisors at the local university. She cultivates relationships with journalists who are from local communities and are able to give her tips or help make connections in communities where she has less access.

Dasgupta recommends searching top scientific journals, databases, and university departments in the fields you report on to find new voices. Add the new people to your contact list for future stories and consider doing informational phone interviews to find out more about them and their research.

Don't stop with just one new source, says Lee. "Go on friend-making missions." A little-tapped resource, she says, is "affinity groups" in STEM, such as the American Indian Science and Engineering Society, the National Society of Black Engineers, and the Society for Advancement of Chicanos/Hispanics and Native Americans in Science. These groups hold regional and national conferences rarely covered by the press.

"You can have a front-row seat to meet professionals of color from a variety of fields and start relationships on the ground," Lee says.

For reporters on deadline, Lee started the National Science & Technology News Service, which links journalists to scientists of color who have experience working with the media. Tweeting them (@NSTNSorg) for a source gets a quick response, she says. Other Twitter sources Lee recommends include:

- Black and STEM: @BlackandSTEM
- Latinas in STEM: @LatinasinSTEM
- Vanguard STEM: @VanguardSTEM
- Women of Color in STEM: @WOCinSTEM
- Minority Postdoc: @MinorityPostdoc

When reporting abroad, reaching out to local journalists (also known as "fixers") and scientists can open doors to diverse voices and perspectives. If you are doing a story on Puerto Rico, can you interview Puerto Rican scientists? What about people from the local communities impacted?

National groups such as the Native American Journalists Association, the National Association of Black Journalists, or the National Association of Hispanic Journalists can also help reporters find new voices, says Generation Justice's Rael. She cautions, though, that connecting with journalist affinity groups only to find more sources may not be well received if it's not reciprocated. Be ready to give back, and "learn to be an ally," Rael says. Being an ally means calling attention to racism as well as being willing to advance the careers of journalists of color, share job opportunities, or take on mentees.

Broaden How You Define "Expert"

Diversifying the voices represented in your stories may also mean expanding your definition of who qualifies as an expert. "People in underrepresented communities have perspectives and experiences vastly different from state or federal officials, scientists, PIOs [public information officers], and industry," says Paskus. They may not be traditional "expert" sources, but they are experts on their own lives and the impact scientific developments have on them. Those perspectives can enrich science stories.

People who aren't used to being contacted by the media may be reluctant to speak with journalists. People from marginalized communities have developed a justified skepticism of the media because of past coverage. Many may wonder why they should talk to another reporter who will likely make them out to be poor, violent, uneducated, and helpless.

"You can't parachute in and expect to get your story or quote," said Antonia Gonzales of *National Native News*, speaking at a workshop she co-led with Gustavus at the annual meeting of the Society of Professional Journalists in April 2016. "You have to build relationships over time."

Gustavus covers health and environment stories in which, as in Paskus's reporting, the people most often impacted are Latino/Hispanic, Native, and low-income communities—communities where she is seen as an outsider. "They are taking a risk in talking with me," says Gustavus. She works to build trust by giving them some control over the interview: taking more time, letting people talk without interrupting, letting them know they can refuse to answer any question they aren't comfortable with. She also shows them her past stories, which in her case focus on more positive narratives: successful community events and programs, for example, or how challenges are overcome. Gustavus also attends at least one community event a month—not looking for stories, but just to get to know people.

Science writers can play a role in changing the media landscape, and perhaps even help increase diversity in science, by making time to find underrepresented voices, challenging the dominant science-media narrative by debunking stereotypes, and examining and talking about biases. Doing so may feel like a small contribution to an overwhelming problem. And it can be uncomfortable to discover one's own weaknesses or to risk offending others in the process of learning. "I think [equity and diversity] gets really complicated and scary for everybody involved, and that's why these conversations end up being so difficult to have," Rael says. "But the more we have these conversations, the more we'll end up having the kind of journalism as well as [equitable] representation that our world and our country deserves to have."

21 How to Find Patient Stories on Social Media

Katherine J. Wu

ON A COOL Texas afternoon in mid-April 2020, Anna Kuchment dashed off an email she figured would never get a response.

Kuchment, a science reporter for the *Dallas Morning News*, was writing a piece about convalescent plasma—a centuries-old treatment that can speed an ailing person's recovery by infusing their blood with plasma from someone who's already recovered from the same disease. Preliminary trials of the blood-based therapy, which relies on infection-fighting antibodies produced by the immune system, indicate that it appears to be effective against some severe cases of COVID-19.

To add depth and heart to her story, Kuchment wanted to feature a patient who'd undergone the procedure and could speak to its benefits. Without that added dimension, "you don't really see how this touches people's lives," she says. But Kuchment's typical go-to for patient sources—local hospitals, where staff were understandably overwhelmed—had so far yielded only dead ends.

So Kuchment did something she'd never tried before: She turned to social media. A quick Facebook search for "convalescent plasma" and "Dallas" turned up several posts seeking plasma for a local man with COVID-19 who'd been on a ventilator for nearly a month.

Kuchment, who didn't fit the bill for a donor, wasn't exactly the target audience for the posts. But she emailed the people behind the calls all the same, expressing her condolences, explaining that she was a journalist, and politely asking if she could be put in touch with the family of the patient described.

It was a long shot. "I wasn't really expecting to hear back," Kuchment says. But almost immediately, a message pinged into her inbox. Within hours, she was in contact with the patient's wife, Jackie Hoffman, who had good news: Her husband, Michael, had received plasma from a donor just days before. It was too early to say for sure, but the treatment appeared to be working—and by week's end, the couple had agreed to share their story.

Kuchment's article, published on April 25, 2020, ended up featuring not just Michael (who is now recovering at home) but also his doctor, Gary Weinstein, and his donor, Leonard Seiple. Featuring the perspectives of all three helped build a complete story—one rich with science, history, and human emotion. "It was all because of Facebook," Kuchment says.

Not all health reporting goes quite this smoothly. But Kuchment's first foray into Facebook sourcing represents what appears to be a growing trend in science writing. Thanks to social media platforms, it's become easier than ever for journalists to connect with sources and amplify voices from tough-to-access, vulnerable, and marginalized communities. But venturing into this virtual territory also raises some complex ethical quandaries about the spaces reporters should and shouldn't invite themselves into—and how to balance timeliness and accuracy with sensitivity and compassion.

Leveraging social media for patient stories isn't something to be taken lightly, says Eli Chen, a science and environment reporter for St. Louis Public Radio. The key, she says, is to remain conscious of the value of the information that's being communicated—to both the reporter and the source, who's inevitably taking a risk in sharing their story.

"What I recommend is to just be human," Chen says. "That's how you should treat all of your sources anyway."

Finding an In

An important part of health reporting includes finding narratives from patients who can make otherwise dry figures and statistics more approachable, says Priska Neely of *Reveal*, who has reported extensively on infant and maternal mortality. As she has written, Black women are three to four times more likely to die in childbirth than their white counterparts. "Without the stories of people who have actually experienced this," she says, "those statistics are easy to look at through a blame lens" that puts Black mothers at fault.

Sometimes, those narratives come to journalists in the form of tips, allowing a reported story to be shaped around them. But in many cases, reporters must do the legwork of identifying and interviewing patients and survivors themselves, starting only with the kernel of a topic. Taylor Knopf, who reports on health care for *North Carolina Health News*, often starts her search by querying her other sources, especially doctors or other local leaders who work with people with specific medical conditions. That avenue alone works about 90 percent of the time, Knopf says.

Patient advocacy groups or lawyers (such as those who specialize in malpractice or personal injury) can be a rich resource as well, Knopf adds. When in-person meetings are a possibility, Knopf has also had luck connecting with patients at conferences, rallies, walks, and other events that center on the condition she's interested in.

Then there's social media. Some journalists have found sources by digging through relevant hashtags and lists on Twitter, or scouring Reddit threads for vocal participants. But most of the reporters interviewed for this piece have had the best success with Facebook, especially when seeking sources with a local connection or an especially rare condition.

When pursuing a story about a group of nursing home workers struggling to cope with the unprecedented conditions brought on by the COVID-19 pandemic, Kaiser Health News reporter Anna Almendrala used Facebook to track down individuals linked to the Magnolia Rehabilitation & Nursing Center in California. She found the profile page of the Riverside County Department of Public Health, where a video clip of a press conference had attracted hundreds of comments, and she messaged dozens of people who appeared to have inside knowledge of the situation. Most never responded, but Almendrala eventually found several individuals willing to go on record, including a few certified nurse assistants and a family member of a nursing home resident.

Boston-based science journalist Elizabeth Preston employed a similar strategy when diving into the world of fecal transplants. After punching in a few keywords into Facebook's search bar, she happened upon a few groups that didn't restrict membership. She joined and quickly introduced herself as a journalist looking for people who may want to share their experience. She also noted that if anyone considered her request inappropriate, they should let her know.

This kind of transparent, respectful approach is key to successfully navigating social media as a journalist—especially when covering sensitive topics like medical conditions, Preston says. "It can feel intrusive if a journalist shows up and [gives the impression that they're] trying to mine your group for personal anecdotes. . . . I was careful to be polite and upfront."

Some groups, like several of the ones Preston found, are fairly lax about their membership and are intended to be virtual portals for discussion. Others, though, are more exclusive, and reporters should tread with caution. Both Preston and Sharon Begley, who reports for *STAT*, recommend steering clear of groups that explicitly bill themselves as support groups for patients or individuals with specific conditions. Many of these groups are also closed and require moderator approval to join—likely in part to keep their members safe. If all other avenues seem blocked, one possible workaround might involve sending a message to the admins and explaining the situation. Some moderators may be open to letting journalists join and post a message, or may even write something on their behalf. Even if indirect, a moderator's endorsement can make all the difference to group members who don't have much experience with the press, or don't otherwise have a way to vet who's asking them for personal information.

Respecting Boundaries

Though what's on social media is often public, that doesn't mean it's always fair game to publish. Almendrala says quoting material from someone's account—including, for instance, embedding a tweet—shouldn't happen "without touching base with that person first."

It's common to encounter patients who are eager to engage on an informal basis, by adding comments or replies to posts from journalists, but wary of going on record. "In general, people are very ready to answer questions on the main feed [in a Facebook group], where everyone is chatting and sharing stories," Preston says. "But if you follow up and ask, 'Can I call you?' sometimes people don't respond, or get uncomfortable."

If the answer to a request for comment is a flat-out no, or radio silence, move on, says Almendrala, who sometimes ends up sending hundreds of messages via various social media platforms before identifying a responsive source. But if someone expresses hesitation or asks questions, rather than simply saying no, it's fine to respectfully ask whether you can address their concerns by providing more information. "I try to reassure [these patients] that I will be sure to quote them accurately and do fact-checking," Preston says. Oftentimes, "people come around when I explain exactly what the process is."

It's important to remember that when seeking out patients on social media, journalists are venturing onto their sources' turf—not the other way around. That requires using a "lighter touch," perhaps more so than usual, Almendrala says.

When in doubt, consider how you'd feel if someone screenshotted the messages you were sending to potential sources, Neely says. "Would it be embarrassing? Would you feel terrible? Just remember that this is a person's story."

If a deadline isn't looming, take the time to get to know your source and build up a rapport—give them a concrete reason to trust a stranger with sensitive information. There's "power in building relationships over time," Neely says.

Begley and others also point out that it's good practice to verify that interviewees are okay with having their full names published, especially if the source doesn't have much experience talking to reporters. "I don't promise to show [copy]," Knopf says. But she often offers to review certain parts of her story with her sources before sending off the final piece. "I'd be leery of putting my own personal medical history out there, and I take that into account."

Sensitivity is also crucial during fact-checking—especially if medical records are involved. Requesting these types of documents isn't always necessary, Preston notes, especially if an individual is just recounting their personal experience with a condition. But in other cases, such as those in which patients are making accusations against hospitals or nursing homes, claims need to be cross-checked with some sort of official record. Here, Neely tries to be as transparent as possible. "Just explain why," she says. Having documentation simply makes it easier for a reporter to write a clear, accurate story.

Keeping Lines of Communication Open

Once you've established something of a beat, "put yourself out there" on social media, and invite patients with stories to find *you*, Neely says. "Have a pinned tweet that says, 'I'm looking for stories about [some topic], here's my email, my DMs are open.' You never know what someone may tell you."

And for many health journalists, the relationships seeded while reporting stories extend beyond the publication date. After wrapping up a recent COVID-19 story, Chen of St. Louis Public Radio stayed in touch with a source whose mother had tested positive for the virus. "I wanted to know if she was doing alright," she says. "Journalists are often seen as extractors . . . who file a story one day and then just move on to the next thing. I really don't want to treat people's information that way."

Jackie Hoffman, the wife of the COVID-19 patient profiled in the piece in the *Dallas Morning News*, appreciated a few things about Kuchment's approach. Kuchment didn't dog her with a deadline, especially given her husband's fragile condition. She liked that Kuchment said the focus of the story would be on the science of convalescent plasma rather than a lens into Michael's personal life; it was also comforting, she says, to know that Michael's doctor had agreed to be interviewed for the piece.

Through it all, Kuchment was kind and cognizant of the uncertainty of the situation, Hoffman says. "That's the most important part—to respect that, for whoever you're talking to, [getting interviewed by a reporter] is the last thing on someone's mind when they're going through a family crisis."

22 Pulling It All Together: Organizing Reporting Notes

Sarah Zhang

IN THE SUMMER of 1966, the legendary writer John McPhee lay, by his own admission, on a picnic table for nearly two weeks straight. He was trying and failing to begin a story about the Pine Barrens, a stretch of undisturbed wilderness in southern New Jersey. He had already interviewed woodlanders and blueberry pickers and fire watchers; he had read books and scientific papers and a doctoral dissertation. "I had assembled enough material to fill a silo," McPhee writes in his book *Draft No. 4*, and he had no idea where to start.

"I do find some solace that even John McPhee has trouble," says Christopher Solomon, recounting the picnic-table story when I asked about his own strategies for organizing notes to begin writing. Solomon is a contributing editor for *Outside* and *Runner's World*, and like almost all the journalists I spoke to, he says he finds this task one of the most difficult aspects of being a writer.

Every journalist has to both report and write, but these two tasks are in some ways opposites. The first is social, the second solitary. And while reporting is a process of accumulation, of filling one's silo with material, the organizing phase of writing is one of excavation, of discarding storylines, facts, and hard-won quotes until only the most essential elements remain. It can feel a bit like working against your past self.

When Memories Are Still Fresh

Accordingly, think of organizing while you report as a favor to your future self. When Erika Hayasaki goes on a reporting trip, she sits down to organize her thoughts every evening. She does so by writing partial scenes and impressions. "I'm just doing that vomit of notes thing, but it's more focused vomit," she says. In 2015, Hayasaki wrote a story for *Pacific Standard* about a doctor with mirror-touch synesthesia, or the ability to literally feel what his patients are feeling. To report that story, she spent a long day in Boston shadowing the doctor. "I was exhausted and I still came back to the hotel and wrote a lot of stuff out," she says. She spent about an hour doing that.

Hayasaki especially wanted to record details that resonated with her emotionally, and that she thought would resonate with a reader, too. She tapes her interviews and takes extensive notes, but the feelings evoked by certain moments or details, she knew, could be fleeting. That day in Boston, she remembered the way one patient clutched a stuffed blue bunny. She remembered the bewildered expression on his face. Later in the evening, she wrote out a scene describing the doctor's interactions with this one patient. It ended up becoming her lede.

Lizzie Johnson, a reporter who covers wildfires for the *San Francisco Chronicle*, is even more systematic about reviewing her notes. After each day of feature reporting, she goes through and color codes her notes with highlighters: yellow for good quotes, blue for key details, and pink for statistics. Then she types up a paragraph or two distilling what happened.

Johnson used this strategy to report a feature that followed two families for a year after they lost homes in the Tubbs Fire. She first met the families while reporting a breaking news story during the fire—she says she always gets phone numbers when she interviews people in the field— and she met them several more times in 2017 and 2018. "It was a really massive project. I had hundreds of pages of word processing notes by the end of the reporting process," says Johnson. On a long enough reporting project, a writer's enemies are time and the fallibility of their own memory. To combat these foes, Johnson had to be organized from the get-go.

Rereading Notes Before Writing

When all the reporting is over but before the writing begins, Johnson, like many writers, rereads all of her notes. She prefers to do this the low-tech way: She prints out her notes and lays them down page by page on the floor of a long hallway.

At this point, Johnson is rereading the notes with an eye toward her story. For example, for her Tubbs Fire piece she had to figure out how to weave stories of the two families together. She began by writing out a timeline of both families and highlighting events that overlapped thematically. (She's big on highlighting, in case you haven't noticed.) These overlapping moments—for example, family escaping from the fire, family debating the future—were used to cut from one family's story to the other's. Johnson had specifically chosen an older couple and a younger couple, and these points of overlap became natural ways to compare and contrast the two.

Other writers swear by specialized software products such as Scrivener and Evernote, which corral interview notes, PDFs, photos, and other research materials into one place. Libby Copeland, a freelance writer for the *Washington Post*, *Slate*, and other publications, puts all of her notes in Scrivener. This program allows users to link their writing to their research—helpful for later fact-checking—and to work with their project in an overview mode, in which it's displayed as index cards on a corkboard. But the best thing about using Scrivener, Copeland says, is that she can search all her files at once. If she needs to find every instance where any source brought up a topic, a single Scrivener search brings them all up.

Solomon, for his part, has begun using a little-known function in Microsoft Word to organize his notes by topic. (He credits Thomas Curwen, a Pulitzer Prize–winning staff writer at the *Los Angeles Times*, for teaching him the trick.) When Solomon sits down to outline, he begins by listing all of the major topics that will go into his story. He goes from A and Z, and if he runs out of letters, he starts going down the alphabet again with AA, BB, CC, and so on.

Then, as he rereads his typed notes, he tags every fact, number, or detail he wants to use according to which topic it addresses—for example, by adding an "A" at the beginning of every paragraph that corresponds with topic A. Then he uses Word's alphabetical sort function on the entire document—essentially sorting all of his notes by topic. (The "Sort" button is found under the Home menu in Word.) To take advantage of the sort algorithm, Solomon warns, you do have to be quite precise with your inputs. For example, he puts a period and two spaces after the topic headings (for example: "A. ") so the sort function can pick out just the tagged paragraphs. He suggests keeping a backup version of unsorted notes, in case something goes awry.

Finally, Solomon takes his alphabetically sorted list of topics and re-arranges them into an outline with pen and paper.

How Much to Outline?

Outlining is the first attempt at structuring a story, and there is no single correct way to do it. Copeland told me she just scribbles a quick outline in her notebook. Solomon, in contrast, calls himself a "religious out-liner." His handwritten outlines can run five pages long, and he might revise his outline several times in the course of a single story. During this process, he also begins sketching out the scenes that will be the structural backbone of his story. Each scene serves a purpose.

Solomon's *New York Times Magazine* profile of a renegade wolf biologist, for example, opens with the scientist bringing a gun to a bar. "It's just an absurd scene," he says, that gets at the scientist's "over-the-topness." Another scene, set at a location where wolf pups were killed, allows Solomon to tell the backstory behind the scientist's paranoia. Solomon also uses quick vignettes that came from driving with the scientist as connective tissue between more fact-heavy sections.

Another way to start an outline is to start with the big idea. Linda Villarosa, a contributing writer for the *New York Times Magazine*, always writes her nut graf first. (A nut "graf" might actually run multiple paragraphs for a long feature.) "Otherwise it's very easy for me to go off topic," she says. The process of writing the nut graf forces her to think through all of the biggest questions: What is the theme of the story? Why does it matter? And why do we care now? "I just bang that out, even though it's not the prettiest thing at first writing," says Villarosa. Only then does she write the rest of her outline, using her nut graf as a guide to avoid straying off topic.

What Doesn't Work

When writers told me about organizing methods that did *not* work for them, the failures seemed to have a common theme: They didn't take out enough extraneous material. Solomon, for example, says he used to take notes on his notes. "I'd end up with 14 pages of notes on my notes," he says. Copeland tried another popular method for one story, putting facts on index cards that she could then shuffle around to try different structures. The problem she ran into, she says, was that committing a fact to an index card made her too attached it. "There was an element of dutifulness," she says. "Once I pulled out the cards, it was almost like I felt obliged to get every piece of reporting on a card." And once it was on a card, she felt obliged to put it in her story. Her first draft for that story ended up far too long, and that dutifully transcribed material had to be cut anyway.

Neither of these methods is inherently bad, of course—some writers even swear by them. (A popularizer of the index card method is no less than John McPhee, whose Pine Barrens piece was ultimately published in the *New Yorker* and then expanded into an acclaimed book.)

But it's okay to try out different methods to find the ones that work with your brain, says Hayasaki. She's known writers who have to begin with an outline and other writers who just start "vomiting" a draft. "It's different for everybody," she says. The key is to find a method that allows you to pare down material and synthesize what remains—rather than organizing simply for the sake of being organized.

23 Gut Check: Working with a Sensitivity Reader

Jane C. Hu

SCIENCE WRITER Kate Horowitz had just landed an assignment she was excited about: a reported essay for the online magazine *Bright Wall/ Dark Room* exploring trauma and recovery in the Guillermo del Toro horror film *Crimson Peak*. The film shows how the movie's three main characters underwent childhood trauma, including physical violence and a mother's death. In her essay, Horowitz probed the characters' universe through the lens of trauma psychology. As she researched and wrote the piece, she strove to represent trauma survivors' challenges through recovery as accurately as possible; she carefully researched trauma, and drew on her own experiences as a trauma survivor. At the same time, she knew that her experiences couldn't represent all survivors', and she was wary of making generalizations.

"Trauma survivors have been through enough—they don't need me to get this wrong, too," says Horowitz.

So Horowitz decided to enlist the services of a sensitivity reader: someone who would review her work with an eye toward accurate representation of marginalized groups. The practice of sensitivity reading originated with fiction writing, and it's still relatively rare in journalism. But working with a sensitivity reader can be an important step toward crafting richer and more accurate stories.

Because it's not built into most publications' editorial process, the decision to enlist a sensitivity reader often falls to writers. Though publications may have some in-house resources that can be helpful—for example, an Asian American editor may be asked to read through a colleague's piece on the use of CRISPR in China—staffers may not be knowledgeable about the particular communities or places a writer covers in a piece. Hiring a reader can help writers have confidence in their piece's framing and details even *before* editors weigh in.

And that's what Horowitz did: For her essay, she worked with a therapist who specializes in trauma, who provided feedback on Horowitz's discussions of current trauma theory and recovery. Horowitz paid $50 for an hour of her reader's time. "It was absolutely worth it," she says. Hiring a reader gave Horowitz peace of mind, not only that she'd accurately represented the science, but also that she'd approached her discussion of survivors' healing with the requisite care.

Like fact-checking, sensitivity reading can help illuminate the truth by avoiding harmful stereotypes or mischaracterizations. Montreal-based reporter Selena Ross says she works with sensitivity readers when writing about cultures or places she's not familiar with, because they provide valuable insight on the big-picture issues in her portrayals. "You want people from that place or culture to read it, respect it, and feel like it's true to reality," she says. Just as she calls experts to explain a technical concept, Ross says, she calls experts with experience participating in a particular community. When she wanted to double-check the cultural nuances of a story that took place in northern Canada, an area unfamiliar to her, she asked a friend who was from a northern Indigenous culture to read it over. "The last stage in my process is calling someone who knows the subject . . . to make sure I framed it right," she says.

Gaius Augustus, a multimedia communicator and visual storyteller who is transgender nonbinary, says they believe that any story that highlights experiences outside the journalist's own ought to be run by a sensitivity reader. Augustus has been on both sides of the process—they've served as a reader for writers working on pieces that include transgender people, and have enlisted sensitivity readers on other topics.

Augustus says that while writing articles about people with cancer, they asked several people to share their opinions on the language they used. People commonly use terms like "cancer survivor" or "cancer patient," or use metaphors of war such as "battling" or "fighting" cancer. Some people take issue with those characterizations, while others find strength in them. "Because I have never had cancer and haven't been close to anyone going through that experience, I wanted to make sure my language was appropriate," says Augustus. In deciding what language to use, they say, they follow a source's lead and use their preferred language.

While many science stories could benefit from sensitivity readers' perspectives, it may not always be logistically or financially possible to enlist a reader. The best way to know whether you should commit to the process is to listen to your gut. Do you feel like you're out of your depth, or worry about wading into controversial territory? That's what drove Horowitz to hire her reader. "I knew I would not be comfortable putting it out there unless someone with more knowledge than me and more grounding in [the trauma] community, research, and practice could review it," she says.

Deciding to Use a Sensitivity Reader

Writers sometimes don't begin to consider the need for a sensitivity reader until a story is nearly done. But the ideal time to consult such a reader is early in the writing and reporting process. The very framing of a story can be problematic, and a sensitivity reader's early feedback might educate writers on key facts, issues, or perspectives to help shape the piece.

For example, in April 2019, *Science* published a feature by journalist Sam Kean about scientists' role in the transatlantic slave trade in the 1700s. The article prompted complaints, in part because it centered the perspective of white scholars, including by quoting sources using the terms "we" and "us" in describing scientists' lack of knowledge about that history. In a published letter to *Science*, ecologist Rae Wynn-Grant, a fellow with the National Geographic Society, wrote that this framing "suggests an underlying assumption that neither the Africans and African-Americans enslaved nor their descendants, who experienced and survived 400+ years of the transatlantic slave trade, were scientists then or are scientists today." She added that although the feature may have intended to raise awareness about white scientists' problematic involvement in the slave trade, the article's language and context "upholds colonial science and white supremacy."

Kean told me he thought Wynn-Grant's letter "raised some fair points," and that "if a sensitivity reader had flagged things, [he] certainly would have been open to altering parts of it." He says he doesn't agree that the article overall promotes white supremacy or absolves those who benefited from the slave trade. In a published response to Wynn-Grant's letter, Kean's editor Tim Appenzeller, *Science*'s news editor, echoed Kean's sentiment, but apologized for the story's language.

Appenzeller told me that the editorial team didn't consult a sensitivity reader but wished they had. "Since then, we have asked for sensitivity reads on stories about similar topics," he says, including for a piece on the archaeology of slavery in the Caribbean.

Other editors might similarly be open to bringing a sensitivity reader on board, and may even have a budget for doing so. Others may not even know it's an option, or may not offer financial support for a sensitivity reader. (None of the journalists I spoke with for this piece had asked an editor or publication to cover the cost of hiring a reader, but it makes sense to at least ask.)

Regardless, writers who are interested in using a sensitivity reader should talk first with their editors. For one thing, some publications have editorial policies that prohibit sharing unpublished copy with outsiders before publication; editors and writers can work together to determine acceptable boundaries for what to share with a reader.

Additionally, working with a sensitivity reader without your editor's knowledge can cause later problems. "If the editor doesn't agree with you or doesn't see [a sensitivity reader] as important, and reverts [suggested] changes, that can sometimes, and often does, reflect on the writer," says Ebonye Gussine Wilkins, a social justice writer and editor who has served as a sensitivity reader, as well as taught workshops and cowritten a booklet about nonfiction sensitivity reading.

Before consulting with a sensitivity reader, it's important to be open to receiving an honest critique and to commit to carefully considering that reader's feedback. Of course, writers are not required to make all (or any!) changes readers suggest, but they should be committed to understanding the roots of any problems their readers identify.

Having a reader review a piece should never be a checkbox or a way to cover bases before publication, but rather a tool to explore what consequences—intended or unintended—a piece might have. And sensitivity reading is not a guarantee that a piece is problem-free. "The point [of such a read] is not to absolve the writer," Gussine Wilkins says. Readers, like writers, bring their own biases and experiences to their work, and may have different ideas about appropriate representations or portrayals. As Augustus notes, "A sensitivity reader should never be considered a representative of a whole minority."

Finding a Sensitivity Reader

In finding the right sensitivity readers for your project, identity is para-mount. Gussine Wilkins recommends enlisting more than one person. Two to three is a good start, but working with "as many [readers] as your budget can allow" will provide valuable insights, especially if readers have intersecting identities. For example, she says, "Disabled people face stigma and ableism in the world, and Black people face a lot of rac-ism and other stigmas in the world, but what is it like . . . when those identities converge?" In that case, hiring both a Black reader and a dis-abled reader could provide valuable perspectives.

A good place to start looking for sensitivity readers is through online directories. Writer and sensitivity reader Renee Harleston keeps a direc-tory of sensitivity readers on her website, Writing Diversely, which lists readers' backgrounds, rates, and reading interests. The group Editors of Color also includes sensitivity readers in its database of journalism pro-fessionals. Writing consulting groups like Salt & Sage and Quiethouse Editing also feature readers-for-hire and list their expertise. Some writ-ers have also successfully sought sensitivity readers from within writing communities or professional organizations they belong to, or by asking for referrals on social media. Successful solicitations include informa-tion about the scope of the work, the perspective a reader might bring to a piece, and the offered rate.

Though some sensitivity readers offer their services through compa-nies and professional listings, others may only sporadically take on work as a reader if approached by writers seeking their insights. A good reader need not have any formal training; lived experience can provide impor-tant perspective.

Horowitz says she looks for someone who is active in the community she's writing about. "I don't want someone who just studied the thing, but [someone] who knows what the current issues are," she says. For in-stance, in her piece on family violence, trauma, and recovery, she was looking for a reader who not only had the expertise to assess facts such as her definition of trauma, but who also had insight into survivors' ex-periences and the ability to weigh in on whether her descriptions of cur-rent treatments and survivors' behaviors were accurate.

While writers sometimes ask trusted friends or colleagues to do a quick review of a piece as an unpaid favor, consider paying your reader for their expertise. After all, reading and commenting on a piece is a type of editing, which is paid work. As Ross points out, people with specific expertise that makes them sought-after sensitivity readers have probably been asked for this type of labor frequently, whether by journalists or others. "If you're finding someone to weigh in on or explain Inuit history and knowledge to you, odds are they've been asked to do that dozens of times in the last year," she says. "They are asked constantly to do free work of this kind," so it's best to offer compensation for their time.

Moreover, sensitivity reading involves emotional labor. As a reader, you're "putting yourself in a vulnerable position," says Horowitz. In reviewing a piece for offensive, problematic, and inaccurate portrayals, she says, readers are "on the front lines . . . the buffer between the author and the reader." Providing constructive feedback in a productive yet direct way takes time and energy.

Working with a Sensitivity Reader

Once you find the right person to work with, establish expectations. Like writer-editor relationships, each writer-reader relationship results in a unique process. If you're a writer seeking a reader's expertise, Augustus recommends knowing what you want out of the collaboration. "Whenever I start a read, I always ask for the type of critique the requester wants. Some people want a general feeling, while others want answers to very specific questions," he says. And it's fine to tell them that you don't want a change in voice or grammar edits, he says.

When Horowitz reached out to the reader she hired for her reported essay, she began the relationship by explaining her project and her goals in hiring a reader. Then she offered an hourly rate, asking the reader if they thought her offer was fair. Though there are no standard rates for sensitivity reading, resources such as the Editorial Freelancers Association's rate list suggest reasonable hourly rates for similar work, like fact-checking or developmental editing, based on what your project entails. Readers' rates are similar to those for fact-checkers; typically, writers can expect to pay $30 to $60 per hour, though some sensitivity readers may charge more.

Sometimes, readers may decline payment—perhaps because they're happy to read as a favor to a writer, or because they feel that accepting payment would be a conflict of interest. If that happens, Ross recommends finding some other way to thank the person. For example, she gifted one reader a magazine subscription and another a home-cooked meal. "In cases where people agree to do you a favor, find a gesture to say thank you," she advises.

In time- and cash-strapped newsrooms, hiring a sensitivity reader might not always be possible (or at least, newsroom leaders may not make it a priority). For freelancers whose clients don't cover the cost of hiring a sensitivity reader, doing so may be cost-prohibitive; and hiring multiple readers could eat up their entire freelance fee. When hiring a sensitivity reader isn't an option, Ross suggests, writers can contact experts who weren't sources for the story to fact-check specific, sensitive details or historical context they're unsure about.

For example, to talk through her work on Indigenous communities, she's reached out to Indigenous advisors at Canadian universities. Numerous professional groups also offer resources to help journalists approach their work with sensitivity.

In some respects, the very fact that sensitivity readers are so often needed reflects a pervasive problem in journalism: a dismal lack of diversity in most newsrooms. Long-term, systemic changes in journalism, including changes in recruitment, hiring, and promotion practices, as well as improved training and professional development opportunities focused on inclusivity and equity, would ensure that newsrooms are better equipped to handle all stories with sensitivity, from the beginning of the editorial process.

In the absence of such systemic change, sensitivity readers can illuminate overlooked issues and provide insights from communities that editorial team members are not active in. "Stories are definitely better off" for doing so, says Augustus, "because it means more accurate and meaningful representations—and it can even give you new story ideas."

24 When Science Reporting Takes an Emotional Toll

Wudan Yan

IN NOVEMBER 2014, Erika Check Hayden left her San Francisco home for Sierra Leone, to report for *Nature* (where she was then a staff reporter) on how Africans and aid organizations were responding to the Ebola outbreak. Emotionally, that reporting would turn out to be unlike any other story she had done. Even when she was back home in San Francisco, Check Hayden found the news "so devastating," she says. "There was a period of time—pretty much every working day—where I would get up, do an interview, read a story, watch a video, or look at a photo essay, and just cry." And although covering the story close-up was mentally and emotionally draining, Check Hayden knew it was important for her to do so.

At an Ebola treatment center in Bo, Sierra Leone's second largest city, Check Hayden met a family—a mother, father, and their children—who were all infected with Ebola. The day Check Hayden was there, the family had just learned their youngest child died. The father seemed to be doing well with the disease, but it wasn't clear whether or not the mother was going to survive—and her child's death was sapping her own will to live.

Check Hayden couldn't help but feel emotionally affected—nor, she says, would denying her response benefit her work. "Being able to admit to yourself and to access the fact that you are really affected by something motivates you to do your job as a journalist and follow the story," she says. The family's heartbreaking circumstances spurred her to follow their story even after she left Bo. (The family's story ended more happily than some: the mother, father, and remaining two children survived.)

The mental health hazards of covering war, terrorism, violence, and other disasters are well known. But science, environmental, and health journalists can also be at risk. Science journalists might experience what clinical psychologists call "vicarious trauma," which refers to the emotions that arise when reporters bear witness to another person's suffering. Covering topics such as chronic or rare illnesses, infectious disease outbreaks, climate change, extinctions, and other environmental crises can evoke anxiety, fear, and guilt, or even trigger post-traumatic stress. "If you're connecting with another suffering human being and you're doing your work well, you can't help but to be moved in ways that are both positive and negative," says Elana Newman, a clinical psychologist and research director for the DART Center on Journalism and Trauma. "Bearing witness is an occupational hazard that most journalists aren't trained for."

Science journalists—like war reporters—need to take such mental health impacts of their work seriously, says Check Hayden, who is now director of the Science Communication Program at the University of California, Santa Cruz. "We are increasingly called upon to cover crises and traumatic events because we have a particular expertise to bring," she says. "As a profession, we need to develop standards and ethical practices around doing crisis reporting. This would include ways to protect the health and safety of reporters, including their mental health."

Lean on Your Support Network

Being alone tends to make any trauma worse. That's why Newman, who researches the impact that reporting on traumatic events can have on reporters' mental health, says the most important factor in mental health and resilience in the face of trauma is social support. She urges journalists on assignment—especially in emotionally difficult terrain—to find ways to avoid being isolated.

That advice resonates with journalist Dan Fagin, who is director of New York University's Science, Health and Environmental Reporting Program and author of the Pulitzer Prize–winning book *Toms River: A Story of Science and Salvation*, which documents a 60-year saga of industrial pollution in a small New Jersey town and the role that pollution played in a cluster of childhood cancers. While reporting the book, he spent long periods of time with families that had experienced trauma "orders of magnitude higher than my own," he recalls. When the emotional stress of bearing witness to that trauma became intense, Fagin says, "I'd talk about reporting problems I'm running into or writing blocks with my wife," who is also a journalist. "I'd also talk to my friends or kids—I'd talk less about the journalism and [more about] how I was reacting as a human. It was helpful for me to process what I was experiencing."

But many reporting assignments pull journalists away from their friends, partners, and peers back home. One way to mitigate isolation while reporting from the field is to collaborate with colleagues, such as other reporters, visual journalists, fixers, or translators.

When journalist Apoorva Mandavilli (who is editor-in-chief of the autism-focused publication *Spectrum* and also freelances on other subjects) went to Bhopal, India, to report on the lingering public health effects of the 1984 gas leak there that killed thousands and injured many more, she brought a photographer friend, Raj Sarma, with her. "That was the best mental health decision I could have made," she says. She and Sarma witnessed people living in extreme poverty and met people who were desperately ill and who had little access to medical help. One morning, they visited a center that cares for children who became disabled as a result of the gas leak. Such experiences were painful to observe, Mandavilli says, but having a colleague with her helped. "In the evenings, we'd go for a drink and process what we saw during the day or talk about something else entirely."

Check Hayden, too, worked with a local photographer when she first arrived in Sierra Leone. "It was so incredibly helpful," she says. "When you're in the field, you can't really get guidance from your editor, so it's helpful to work with someone else who is knowledgeable and professional."

When you *can* get in touch with them, though, editors can be an important source of support. Not only can an editor's fresh pair of eyes provide focus and clarity to a story, or tone down emotions or details that might not serve readers, they can also be an important piece of the social network that reporters need.

"Editors don't need to be therapists," says Newman. "But they can employ the same reporting skills as they would to a story, to just ask a reporter, 'Hey, that was a tough story. What was it like being in the field?' We're not advocating for editors to do anything therapeutic, other than having an informed conversation." (The DART Center offers several resources that editors can employ in their newsrooms.)

Pace Your Reporting and Take Breaks

While on assignment, reporters can feel pressured to report as much as possible in the time available. But taking breaks from reporting can benefit a journalist's mental health.

When reporting emotionally stressful stories, Mandavilli deliberately builds in time "cushions" at the end of the day. When she went to London to meet the protagonist of her *Spectrum* story "The Lost Girls," which investigated the experiences of girls and women with autism, she planned time to see friends and family after she finished her reporting. "This strategy is contrary to advice people give you, which is to go put all your thoughts and impressions at the end of the day down on paper," Mandavilli notes. "I could only do that for an hour before I had to leave my hotel room and talk about something completely different."

In 2016, I spent three months investigating (for *Nature*, PRI's *The World*, and *STAT*) whether palm oil from Southeast Asia can be produced in an environmentally and socially sustainable manner. After just a week of reporting from the heart of Indonesia's palm oil–producing province, I noticed the mental toll that the work had on me. Constantly looking at land that had been plundered for the needs of the world and listening to the stories of workers who were exploited, I felt uncharacteristically sapped of energy and had a hard time focusing on my work.

Altogether, I reported in three countries, for about three weeks each. After I completed my work in one country, I took a week off to meet up with friends in different parts of Thailand to disconnect from my reporting, before jumping back in. These self-imposed breaks—in settings that were so distinct from the hazy cities and palm oil plantations I was reporting in—helped me recharge so that when it was time to report again, I was excited to do so.

Look for Positive Stories

Environmental journalist John Platt is no stranger to the emotional cost of covering a beat that often lends itself to dismal stories. Now editor-in-chief of the *Revelator*, which does investigative reporting on the environment, wildlife, and other related topics, Platt has written about ex-

When to Seek Help

While it's normal to feel sad or emotional when reporting on communities or ecosystems under stress, Elana Newman of the DART Center on Journalism and Trauma advises reporters to notice if their negative reactions—such as dreading going to work, avoiding assignments, being late, turning to drugs or alcohol—become repetitive.

"When you see these reactions taking over your life in ways that are not productive, it's time to seek professional help," says Newman. She says post-traumatic stress disorder and depression are two of the major mental health issues that affect journalists. But both are readily treatable, especially in the earlier stages.

"There's a lot of stigma associated with these conditions, and I really don't think there should be," she says. At the first sign of distress, she advises journalists to speak with someone they trust—and it doesn't have to be a professional therapist—to think through their feelings.

tinction for more than a decade. "The collective impact of reporting on extinction," he says, "is the death of a thousand paper cuts."

To slow the bleeding, Platt mixes his reporting on extinction with freelance reporting on the technology beat. He has recently written about teenagers who are designing self-cooling solar cells and about devices that can detect the onset of Parkinson's disease. "Being able to write about something completely different that is inherently positive has been helpful," he says. He also urges reporters to seek out a fun or hopeful story on the environmental beat from time to time. "If you can tell a positive story that shows your readers that not everything is terrible, you remind yourself of the same thing."

When I was working on my palm oil stories, I tried to not spend long, unbroken spells of time reporting on depressing plantation conditions— the kind of work I thought was exerting the greatest toll on my mental health. I alternated weeks of interviewing people who were suffering from their work on plantations with weeks in which I would speak to NGO leaders, government officials, policy makers, or scientists—people who were thinking about solutions in the palm oil industry and had a more optimistic mindset.

Do Emotions Have a Place in the Story?

When it comes time to sit down and write a story—particularly one that was mentally and emotionally intense to report—strong waves of emotions might resurface and leak into a reporter's writing.

In a feature story for *Nature* about how the Ebola outbreak in Sierra Leone affected the ability of a hospital to continue its research, Check Hayden took a solemn tone. Her choice of language in the story, she says, reflected the emotions that her sources—and she herself—felt in the wake of the deaths of many of their colleagues.

Mandavilli says she tries not to censor herself when she writes her first draft. When she was writing "The Lost Girls" for *Spectrum*, Mandavilli says, her editor, Kat McGowan, highlighted certain details and asked why they mattered for the story. "I think a really good editor can connect certain facts back to the main point of the article, rather than having the reader potentially going down a really sad journey," Mandavilli says.

Some reporters worry that allowing their own emotional experience to seep into their stories reflects a failure of journalistic objectivity. But journalists' emotional responses to the situations they observe are a legitimate part of the factual record. Fagin says communicating your own reactions can be a service to your audience. After all, the most emotionally resonant stories are often the ones that will stick with readers.

However, Fagin cautions that you should "make distinctions between your analysis of facts and evidence—which should be dispassionate—and your descriptions of characters, which can be passionate."

When Platt started writing about endangered species, he tried to infuse some humor into his writing—to bring levity to an otherwise serious topic by making fun of the people or situations that were causing extinctions. But as time has gone on, his writing about extinction has gotten more and more "deadly serious," Platt says. "I think that's the best way to address it—to be flat-out serious, respectful, and approach the topic with the right amount of anger and righteous journalistic fury while being fair, honest, and truthful in your reporting."

25 A Conversation with Annie Waldman on "How Hospitals Are Failing Black Mothers"

Tasneem Raja

WE ALL KNOW—or need to know—that race intersects with every facet of American life, from the mundane to the momentous. Where you sleep at night, what you eat, where you send your kids to school, who you're friends with: Whether you realize it or not, decades of racial segregation and inequitable social design have played an often-invisible hand in shaping the options available to you. Options are at the heart of Annie Waldman's data-driven investigation into maternal harm at hospitals that disproportionately serve Black mothers. When a Black woman in America goes into labor, which hospital she chooses for her delivery—or where the ambulance takes her, sometimes against her wishes—can determine whether she and her baby will leave the hospital together and alive.

Waldman's "How Hospitals Are Failing Black Mothers," published on December 27, 2017, as part of *ProPublica*'s award-winning Lost Mothers series, makes the invisible visible, and it does so with data. Since race, class, geography, and poverty are so deeply intertwined in America, it can be hard to tell where one strand ends and another begins. That's why *ProPublica* started with a large dataset—67,000 cases of women who experienced serious complications in the delivery room—from three diverse states: New York, Florida, and Illinois. Waldman's team identified hospitals in those states that are "Black-serving," meaning higher proportions of their maternal patients are Black. Even when accounting for factors like education, income, and overall health, they found that women who deliver at "Black-serving" hospitals experience higher rates of serious complications. According to their analysis, "Black mothers who are college-educated fare worse than women of all other races who never finished high school. Obese women of all races do better than Black women who are of normal weight. And Black women in the wealthiest neighborhoods do worse than white, Hispanic and Asian mothers in the poorest ones."

Of course, each data point represents someone's reality: a mother-to-be whose long-anticipated delivery turned into a preventable nightmare; a family hollowed by grief; a child denied the chance to grow up with a healthy mother—or at all. The best data journalists spend as much time encouraging people to speak as they do staring at spreadsheets. Waldman spent weeks traveling to meet women who experienced the high-risk complications seen in this dataset, women who survived but lost their babies, or the loved ones left behind when women died. Their stories and their questions—Why did this happen? Why didn't anyone stop it? Why does this keep happening?—take this problem out of the realm of academic journals and, hopefully, into the realm of public outrage and prevention.

Waldman's investigative report expertly weaves numbers and testimony, and shares a methodology for other reporters across the country to replicate her results and identify harmful hospitals in their own backyards. It also points to field-tested delivery room protocols that are proven to work in high-risk births, and could be adopted by more hospitals. Finally, Waldman counters the notion that women who are living with poverty, lower rates of education, and preexisting health complications should not have the same access to safe, healthy maternal care as women with greater resources. Here, she tells the story behind the story.

TASNEEM RAJA: Let's start at the beginning. What made you decide to take on this story, and why now?

ANNIE WALDMAN: *ProPublica* had been working on what we called our Lost Mothers series for several months at the point that I came onto the project. My expertise is really looking at data, analyzing huge datasets that normally are very unwieldy for traditional reporters, and integrating those datasets into stories. I was brought on about halfway through the project in order to provide a deeper data understanding of maternal mortality and how hospitals played a role in the deaths of mothers.

Earlier on, we had started collecting data from a number of different hospitals across the country. The idea was to look at the data and start to see what kinds of patterns we could find. We had a theory, of course, before we started diving into the data, that if you look at hospitals that serve a high percentage of mothers who are mothers of color, you might see higher complication rates. After we analyzed the data, that's exactly what we found.

There have been a number of stories looking at maternal mortality outside the United States, specifically in less industrialized countries or less developed countries. That lens hadn't been put on the United States, despite the fact that if you look at the statistics, they're quite troubling.

In the US, we have an incredibly high complication rate, and an incredibly high number of women who die in childbirth. Every single year, 700 to 900 women die from causes related to pregnancy and childbirth. There are significantly more who experience severe complications—about 50,000 women annually. In the US, there are significantly higher rates of maternal mortality than in other more industrialized or wealthier nations around the world. Not only that, but the rate of mortality has risen over the past decade. So this was a surprising statistic for us to look at.

Now, if you unpack that data, you see that Black women are more affected by severe complications of childbirth and maternal deaths than women of other races. Not only journalists but also academics and other medical experts have really been trying to understand why this is the case. Why are the outcomes for Black women leading the maternal mortality rate in the United States? That's one of the drivers for why we wanted to look at this project.

I want to come back to race and unpack that more. But taking a step back, you examined nearly 70,000 cases involving hemorrhages in Florida, New York, and Illinois in 2014 and 2015. Tell me why you focused on those states and those years. Was it simply a matter of what was available?

When you choose data for a project, it's always a combination of a number of things. For us, it's what was available, but it's also that we wanted to look at urban centers, or areas where there was a higher percentage of Black mothers. When we looked at demographic data, it seemed that those three areas were going to provide us with enough data points that we wouldn't be looking at small numbers and trying to extrapolate statements from that.

I'm always interested in hearing more about the paper trail. Can you tell me exactly how you got access to the medical records you analyzed?

New York State actually allows a lot of this data to be publicly available; it's a dream of mine that all of this data could be publicly available across the country.

We were looking at in-patient discharge records of hospital stays. Most states collect this data as part of their billing practices. We wanted to understand whether we could use this data to also look at things like complications in maternal deaths. For hemorrhages, for example, how much blood is actually used in transfusions is an indicator.

We also looked at other data in this set including patient diagnoses, race, age, and whether a patient had high-risk characteristics—for example, diabetes or high blood pressure. Even though there are small differences in how each state records this data, we do believe that it is a comparable dataset.

You write that it's long established that Black women fare worse in pregnancy and childbirth, and that there is growing evidence that race and racial segregation are driving these disparities. For people who pay attention to these things, this really isn't in dispute at this point. But you also talk about the fact that individual hospitals aren't named in studies, and so from those studies, we don't get to learn or understand the specific stories of real women and the complications and outcomes they faced. Did you think of your story as serving as a supplement to the medical research that's out there, a way of filling the gaps?

That's exactly right. There are incredible researchers who are out there describing these trends. But as a reporter, it's not enough to just describe the fact that Black women are three to four times more likely to die in childbirth than white mothers. We wanted to say what hospitals are contributing to these figures, and where they are. It's not just about establishing that there is a problem. It's about revealing where these problems are, and potentially what issues in those areas could be changed so there could be real impact and change.

That's why, when we looked at these datasets, we wanted to find individual hospitals where these problems were very severe. That's why we turned our eye to SUNY Downstate and two others in New York. We felt comfortable naming these hospitals as places where a lot of changes could be made to improve the safety and care of women, and particularly women of color.

You introduce us at the start of your piece to Dacheca Fleurimond. She's a Black woman who delivered her twin boys at SUNY Downstate. She didn't survive her stay. Tell me how you found out about her story, and why you decided to focus on her. I'm sure you learned about a lot of similar tragic cases. Why did this story stand out to you?

I'm a self-professed data nerd, but it's never enough just to have the numbers in your story. You have to have the humans behind it. Oftentimes, when you're working with data, as a health care reporter or a science reporter, or even as an academic, you forget that each data point is a human. Behind each human there are stories and relationships and connections that should be reported on and told. There's narrative there that could illustrate something for the larger audience.

We were already looking at SUNY Downstate as a hospital that had a high complication rate. I went down to the courthouse and looked at dozens of malpractice lawsuits against SUNY Downstate and some other hospitals that had high complication rates. We were trying to find individuals with stories that really illustrated what we were finding in the data.

Coincidentally, we actually had a whistleblower from SUNY Downstate who reached out to us, somebody on the nursing staff who said that they had witnessed a woman perish on the delivery ward. She thought it was due to failures of the hospital. We had reached out to a number of midwife organizations and doula organizations and asked them to put out emails to their colleagues, saying we're reporting and we want more information on this. There was a huge social push with our reporting, trying to get people to contact us who had stories to share. So even though you could call it a coincidence, I don't necessarily think it was. There were a lot of problems with this hospital, and it was a matter of making sure the right people understood that we were reporting on these issues.

We communicated with that whistleblower. I met with them a number of times, and was able to connect with other people who were part of the hospital and either were there the day Dacheca passed away or knew other people who were there. From there, once I had her name, I reached out to Dacheca's family members. It was an incredibly difficult process to gain their trust. That can be lost as well when we focus so much on data reporting. We can get these numbers easily by putting out public records requests, and we forget that there's this whole other step of making sure that the people you're reporting on, who are represented in these numbers, actually trust you when you show up at their doorstep. That takes a lot of time and a lot of honesty about what you're doing, and transparency, to make sure that they understand that what you're doing is trying to hold the place where this very serious and sensitive thing happened accountable.

I spent many weeks meeting with family members. Dacheca's family was based in Brooklyn, but they also had family members in central Pennsylvania. I drove out there several times to gain their trust. I had been in touch with the seasoned malpractice lawyer who had taken the case on. I had to gain the trust of the lawyer as well. I didn't know at first whether she was going to share medical records with me. It was a lot of listening and a lot of patience, as well as crunching those numbers.

Dacheca Fleurimond consciously made the choice to deliver her baby at a so-called "Black-serving" hospital, because she felt more comfortable there. But you also tell the story of Merowe Nubyahn, who begged not to be taken to Brookdale University Hospital Medical Center because she knew it had a bad reputation. She ended up dealing with absolutely nightmarish harm to her own body—and her baby died in the hospital not long after delivery. Was it important to you to include a story where a patient didn't choose to have their baby at a particular hospital, and in fact actively fought against it?

It's easy to think, "Well, if these numbers are so bad, why would women keep going to these hospitals?" It almost takes agency away from the women, right? It's assuming that they don't look up the same numbers that other people do, and that's absolutely not true.

There is a whisper network among women about where to deliver your babies and where not to. I think it's important to illustrate that a lot of women—even if they know these hospitals aren't necessarily going to provide them with the care they deserve—don't necessarily have a choice, because of the way our health care system is set up. There are public hospitals in areas that, as with our residential system or our school system, are incredibly segregated. We don't put enough care into hospitals that predominantly serve communities of color.

At several points in your story, your analysis seems to almost be trying to err on the side of caution when it comes to identifying the extent to which race is a driving factor behind these higher rates of significant complications. For example, you limit your patient pool only to mothers who are of average birthing age. You note that Black mothers in New York City who are college-educated fare worse than women of all other races who never finished high school. Were you anticipating pushback along those lines? Did you feel you had to cross every "t" to show that race really is at the heart of these disparities?

Some people will want to explain away certain facts or trends that they see in the data through correlating characteristics. For example, if you look at patients who have diabetes, oftentimes you'll see a higher rate in women of color, specifically African American women. So, a lot of people then try to tie that, as a correlating factor, to maternal complications.

I think it's really important to address those factors, of course. But I talked with numerous people who work in hospital safety at a number of hospitals around the country, and in New York City, and they said that at the end of the day, these people are still showing up in your emergency rooms. These patients are still showing up to deliver their babies at your hospitals. The fact that they're walking into your hospitals with complicating factors is not an excuse for poor care. If doctors are aware of the fact that a patient has characteristics that might put them in harm's way more than other women of different races—like high blood pressure, diabetes, obesity, or anything like that—then doctors have to make sure that they're doing things to protect women from those higher-risk factors.

That's what was so shocking about what happened to Dacheca Fleurimond. They recognized that she was an incredibly high-risk patient. She ticked off a number of factors. Despite that, they did not provide her with the extra care that she not only required, but that she deserved. That's why she had that embolism. In situations like that, you can run a regression with the data and you can explain it away by saying two things are correlated. But that doesn't excuse the fact that, as a society, we have to make sure that we provide care for individuals with high-risk characteristics.

I know from my own work in data reporting that states like New York, Illinois, and Florida tend to have rich and robust data to draw from. For someone who's interested in doing a similar analysis in a less open state, a smaller locale, or a more rural region—for example, I report out of East Texas now—how would you recommend trying to tackle reporting on this issue?

We provided a methodology for how we measured complications. The idea behind why we produce these methodologies at *ProPublica* is to provide a potential template for other reporters who are out there— whether national or local—to reproduce our work. It's really important that we actually reveal the insides of our reporting and how we came to our conclusions.

In-patient data on hospital stays aren't required only in the areas that we looked at; it is possible to get this data for other states because of Medicaid and Medicare laws. There is pressure from the federal government on state governments to provide this information so that people understand billing practices across the country. If you go to your state and you ask for in-patient discharge records of hospital stays, it is possible to get this data. Then, if you reach out to the California Maternal Quality Care Collaborative, it's possible to get the metric that we used (which is hyperlinked in my methodology) and apply similar techniques in order to understand what the complication rate might be for a certain hospital in your state.

Let's talk timeline. How long did this reporting take you, and does the end result map neatly with the story and the analysis that you expected to undertake, or did you end up going in surprising directions?

Surprisingly, it was actually short—for us—meaning that sometimes we can spend over a year on a single investigative piece. This piece took me about three months, a relatively short amount of time for how much data analysis went into it, finding sources, confirming things, getting records, and following up with whistleblowers.

I did have a thesis going into this, but I didn't expect to connect with so many actual cases of harm. Especially with maternal harm—but really with any kind of medical harm—people are frequently embarrassed about what happened to them, or feel that it was very private grief that they experienced.

When our social team was reaching out to women across the country on Facebook, Twitter, and other social platforms to try and get them to talk about the harm they had experienced, it was a very difficult task. Frequently, when you do an interview about a woman who has died in childbirth, it's reliving that trauma for that family. When we started looking at this one hospital, SUNY Downstate, I was surprised by how many women actually felt comfortable coming out to talk, and how quickly they realized the importance of sharing their story for furthering the accountability in their case.

You include context and insights from medical experts in OB/ GYN and maternal health from all over the country. You must have anticipated that you weren't going to get a lot from the top-level administrators and doctors at the hospitals you focus on in the piece. Did you feel it was especially important to assemble a squad of impartial outside experts for that reason?

Exactly. I was assuming that most of the hospitals that we were looking at would probably not go on the record to talk to us. I knew it was important to have those outside experts, who could say, "This is what normally happens, this is what should happen, and this is what shouldn't happen."

As I was doing my data analysis, looking at the hospitals, and reaching out to families, I was also reading as many studies as I possibly could around the themes I was looking at—on pulmonary embolism, on hemorrhage and childbirth—and looking at the researchers who had already looked at similar numbers. Then I reached out to all of them and said, "Even if it's off-the-record or on background, would you be willing to comment on this case I'm looking at?" I had the opportunity to share medical records of individuals that we profiled in our piece with our experts, so that I wasn't misinterpreting anything.

Investigative journalists always have to be careful about confirmation bias. Our lens is to always be looking for harm, always looking for areas where we can provide accountability. We have to make sure that somebody else who's objective, who's not doing the project with us, can tell us, "Yes, that is indeed harm, and that's something that shouldn't be done."

PART 4

How Do You Tell Your Story?

NOW THAT YOU'RE equipped with a notebook full of colorful quotes, jaw-dropping facts, and spellbinding anecdotes, it's time to write your story.

The obvious thing would be to sit down and try to tap out the first sentence of your story—but don't do that. You might also be tempted to trawl through your notebook and plop each compelling quote or vivid scene into a document and then construct your story based on them. But don't do that either. Instead, your first step is to figure out what this story is really *about*.

You probably had a pretty good idea of that when you started reporting, but you know far more now than you did then, and your understanding of the story may have evolved. Plus, you've acquired all these complexities and cool details and gone down side tracks that might actually be important. The key to good writing is to first synthesize what you've learned, so you see which parts of your reporting are most important and which parts are distractions (even if sometimes alluring ones).

With a firm hold on what you want to say, you can start putting words on the page. Learning to write well requires learning to read well, so that you build up a storehouse of ideas for how to approach different parts of a story. What seduces readers into a story? What guides them to grapple with the heart of the issue and keeps them from getting confused along the way? How do skilled writers deploy humor, or unspool a mystery, or create the sense of tension that keeps their readers wanting more? What kinds of details and quotes might you sprinkle into a story to give it depth and resonance, or to help build your case? How do you avoid using language that may harm marginalized communities, especially if you're not part of that community and are unfamiliar with how seemingly minor word choices and common metaphors can wound? And how can you bring a story to a ringing conclusion that resonates with the themes of the story to thrum in the minds and hearts of readers?

Oddly enough, even the best writers usually find writing . . . hard. Finding a tidy path through that snarled thicket of information you've accumulated is enormously challenging. And worse, the result of all that effort is almost always a first draft that kind of, well, stinks. Instead of intriguing, enraging, or delighting readers the way you had envisioned from the beginning, it sits lifelessly on your computer screen, an unmoving blob. A blob that can squash your self-confidence.

The only way to achieve the compelling work you're aiming for is through rewriting (and more rewriting). The good news is that this revision work brings the pleasure of watching that crummy first draft gradually get better. More organized. Sharper. More sparkling. Most of all, it turns it into something that will meet the needs of your readers, so that that spark of curiosity that started this whole project can jump from your brain to theirs.

Siri Carpenter

26 Good Beginnings: How to Write a Lede Your Editor and Your Readers Will Love

Robin Meadows

WELL BEGUN IS half done. Scratch that—no clichés. Um. Just as breakfast is the most important meal of the day, the beginning is the most important part of a story. Snore. And that first bit isn't even true. But the second part is. We've all heard the advice:

"I urge you not to count on the reader to stick around. Readers want to know—very soon—what's in it for them," William Zinsser wrote in *On Writing Well*, his classic guide to nonfiction writing. "Therefore your lead must capture the reader immediately and force him to keep reading."

"A good lead beckons and invites. It informs, attracts, and entices," says Chip Scanlan—of the Poynter Institute—in "The Power of Leads."

"It's got to deliver on what you promise," says *New Yorker* contributor John McPhee of ledes in an interview with the *Paris Review*. "It should shine like a flashlight down through the piece."

How do you write a lede that does all this? Here are four editors on ledes they love and why they love them, and the writers on how they did it.

Make the Reader Smile

"*Interstellar*'s True Black Hole Too Confusing"

By Jacob Aron for *New Scientist*

LEDE: Even black holes wear makeup in Hollywood. Last year's hit film *Interstellar* used real scientific equations to depict what happens when a team of space farers venture near a supermassive black hole. Now, a joint paper published in the journal *Classical and Quantum Gravity* from the movie's visual effects team and scientific consultant reveal that the real black hole was deemed too confusing for audiences, and some of the science had to be toned down.

THE EDITOR'S TAKE: "This is my favorite lede of any story I've worked on in the past year," says Lisa Grossman, physical sciences news editor at *New Scientist.* "It's hilarious, evocative, and gives you an accurate sense of what the story is about all at the same time."

The lede should get to the point immediately, Grossman says, with the news nearly always in the first sentence. She also likes humor: "The perfect lede ideally first makes you smile, and then makes you keep reading."

THE WRITER'S TAKE: The best ledes give you a taste of the story but leave you wanting to know more, says Jacob Aron, physical sciences reporter at *New Scientist.* Inspiration for a good first line often comes after work, when he's done reporting but hasn't started writing. "I'll have stories from the day before buzzing around in my head, planning out different structures, and something will pop into my head," he says. "Sometimes I'll be just falling asleep when something occurs to me, so I blearily grab for my phone and email it to myself."

Aron didn't have time to wait for inspiration for his *Interstellar* lede, though, because the story had a short embargo. Luckily, an idea popped into his head on the spot. While the press release pitched the "Hollywood does real physics" angle, when Aron read the paper he realized that the black hole's original depiction had been even more real—but so strange the director worried it would leave viewers in the dark.

"I immediately loved this, as it just seemed so typically Hollywood that even the supposedly most realistic depiction of a black hole ever wasn't safe," Aron says. "And then I got an image of this monstrous singularity in the fabric of the universe sitting in a trailer, doing its makeup before being called onto the set."

Leave the Reader Hanging

"Mysterious Fairy Circles Are 'Alive'"

By Rachel Nuwer for *Science*

LEDE: Walter Tschinkel may not have solved the mystery of the fairy circles, but he can tell you that they're alive. Tens of thousands of the formations—bare patches of soil, 2 to 12 meters in diameter—freckle grasslands from southern Angola to northern South Africa, their perimeters often marked by a tall fringe of grass. Locals say they're the footprints of the gods. Scientists have thrown their hands up in the air. But now Tschinkel, a biologist at Florida State University in Tallahassee, has discovered something no one else has.

THE EDITOR'S TAKE: This lede has everything David Grimm wants. "I'm looking for a few things in a lede: an attention-grabbing first sentence; a clear, concise, and user-friendly summary of the new finding; and a sense of why the average reader should care," says Grimm, online news editor at *Science*.

Grimm also likes ledes that are "lean and mean." He favors short, snappy sentences that are free of jargon and information that isn't necessary to hook the reader, such as the journal name and research institution.

THE WRITER'S TAKE: Freelancer Rachel Nuwer learned how to write ledes by working with her editors, who "can expedite the process of discovering what differentiates a mediocre lede from an exceptional one." Her experience is common. "I've seen and rewritten thousands of ledes," Grimm says.

Nuwer wrote this fairy-circle story early in her career, and collaborating with Grimm taught her to structure ledes as mini-anecdotes. Her fairy-circle lede introduces a character (biologist Walter Tschinkel), includes the news (fairy circles are alive), defines fairy circles, and—crucially—leaves readers wanting to know more.

Ledes should "present some form of tension, mystery, surprise, or challenge," Nuwer says.

Collaborating with another editor, Richard Fisher at the BBC, taught her that ledes in longer stories should tease the reader—giving bits of tantalizing information early on, but not the whole story. "Regardless of the length of the story, however, the more compelling the lede, the better the chances that the reader will read on," Nuwer says.

Find Common Ground with the Reader

"How to Teach Old Ears New Tricks"

By Gabriel Wyner for *Scientific American Mind*

LEDE: "Hi! I'm Gabe. What's your name?"

"Seung-heon. Nice to meet you, Gabe."

Uh-oh.

"Sorry, I missed that. What's your name again?"

"Seung-heon."

This is bad.

"Sung-hon?"

"Seung-heon. It's okay—just call me Jerry. Everyone does."

I hate it when this happens. I have every intention of learning this person's name, and my brain is simply not cooperating. I can't seem to hear what he's saying, I can't pronounce it correctly, and there's no way I'm going to remember it for more than five seconds.

THE EDITOR'S TAKE: "This is one of my favorite ledes of all time," says Karen Schrock Simring, news and letters editor at *Scientific American Mind*. "It's instantly relatable—we've all been there—and it's an instant scene." She adds that starting with a short dialogue is a lot more fun than just describing this predicament ("Remember the last time you met someone with an unusual foreign name and you just couldn't hear it properly?").

Simring says the best ledes are emotionally evocative and show rather than tell, giving readers an instant mini-experience without telling them what they should be feeling by using words like *tragic, staggering,* or *amazing.* She also likes ledes that introduce the story's main points effortlessly. "It's lovely to read an introductory section of an article and be totally sucked in—so engrossed that you don't even realize you've just hit the nut graf or moved into the meat of the article," she says.

THE WRITER'S TAKE: A good lede connects readers with their own experiences in a funny or provocative way, says Gabriel Wyner, author of *Fluent Forever*, a science-based guide to learning new languages. While distinguishing sounds in foreign languages is not a universal frustration, many of us have had trouble with foreign names. "My goal was to evoke a memory of this sort of struggle," he says, helping readers empathize with people who struggle in the same way while learning a foreign language.

Identifying a part of your story that readers can understand experientially shows them why you care about it. "If you can find that connection, then you can bring them in and have them share in your own excitement," Wyner says.

Make a Promise and Establish Tension

"Fragile Russian Wilderness"

By David Quammen for *National Geographic*

LEDE: Some places on this planet are so wondrous, and so frangible, that maybe we just shouldn't go there.

Maybe we should leave them alone and appreciate them from afar. Send a delegated observer who will absorb much, walk lightly, and report back as Neil Armstrong did from the moon—and let the rest of us stay home. That paradox applies to Kronotsky Zapovednik, a remote nature reserve on the east side of Russia's Kamchatka Peninsula, along the Pacific coast a thousand miles north of Japan. It's a splendorous landscape, dynamic and rich, tumultuous and delicate, encompassing 2.8 million acres of volcanic mountains and forest and tundra and river bottoms as well as more than 700 brown bears, thickets of Siberian dwarf pine (with edible nuts for the bears) and relict "graceful" fir (*Abies sachalinensis*) left in the wake of Pleistocene glaciers, a major rookery of Steller sea lions on the coast, a population of kokanee salmon in Kronotskoye Lake, along with sea-run salmon and steelhead in the rivers, eagles and gyrfalcons and wolverines and many other species—terrain altogether too good to be a mere destination. With so much to offer, so much at stake, so much that can be quickly damaged but (because of the high latitudes, the slow growth of plants, the intricacies of its geothermal underpinnings, the specialness of its ecosystems, the delicacy of its topographic repose) not quickly repaired, does Kronotsky need people, even as visitors? I raise this question, acutely aware that it may sound hypocritical, or anyway inconsistent, given that I've recently left my own boot prints in Kronotsky's yielding crust.

THE EDITOR'S TAKE: "This lede has it all," says Tim De Chant, senior digital editor at *NOVA* and editor of *NOVA Next*. "There's a hook, a story, and some familiarity, but ultimately there's something unsettled and unfinished about it. We want to read more."

The first sentence hooks us. "With just two words—'wondrous' and 'frangible'—Quammen has us thinking, 'Wherever this place is, I really have to see it,'" De Chant says. "But then he pulls the rug out from under us. The sentence is enticing but leaves us wanting."

Familiarity comes from the image of Neil Armstrong walking on the moon as everyone else watched from their living rooms. Next comes the story of Kronotsky Zapovednik and Quammen second-guessing his exploration of this fragile, remote nature reserve. "Again, he's pulling the rug out from under us," De Chant says.

The best ledes "stand alone as their own tiny story," De Chant says. "Not necessarily as a summary of the story to come, but a passage with a beginning, middle, and end." And the end of Quammen's lede is also the beginning of the story of his visit to Kronotsky Zapovednik, making a seamless transition for launching into the bigger story of the reserve.

THE WRITER'S TAKE: David Quammen learned to write ledes by reading the best—literary writers like Samuel Beckett, William Faulkner, and Albert Camus ("My mother died today. Or maybe it was yesterday . . ."), and nonfiction writers like Loren Eiseley, John McPhee, and Robert Ardrey ("Not in innocence, and not in Asia, was mankind born").

These writers taught him "to grab the reader, hold the reader, give the reader immediate and potent reasons to proceed past the first sampled line," Quammen says. "I came to understand, gradually and unconsciously, that the opening of a piece of nonfiction has to do two things, and do them very quickly: offer a promise and establish tension."

Inspiration for Quammen's lede to this *National Geographic* story came from his reaction to hiking in the reserve: An hour or so into the hike, he was appalled at the damage his boots had done to the reserve's delicate crust and mossy banks. "That's where the lede of this Kronotsky piece begins," he says. "It incorporates my sense of the mandate to offer a promise of natural wonders and establish tension."

*

STILL WANT MORE? In April, Roy Peter Clark of the Poynter Institute published his picks for the best ledes amongst this year's Pulitzer Prize winners. Here is my favorite:

"Dreams Die in Drought"

By Diana Marcum for the *Los Angeles Times*

LEDE: The two fieldworkers scraped hoes over weeds that weren't there. "Let us pretend we see many weeds," Francisco Galvez told his friend Rafael. That way, maybe they'd get a full week's work.

ENDINGS OFTEN ECHO beginnings—but that's the first time I've ever used a lede in the kicker.

27 **Nailing the Nut Graf**

Tina Casagrand Foss

WRITING A NUT graf can feel like showing your work on a math test or stopping at traffic lights when no one's around. Ask journalists about how they constructed a nut graf, and some might actively avoid the question: At least three of the journalists I interviewed for this article talked first about characters and moments in the lead paragraphs. When the conversation turned back to the nut, they said, essentially: "Oh, right. That." Writers seem to discredit nuts because they seem so painfully obvious, so heavy-handed, or so seriously lacking in soul.

It doesn't have to be this way. With a little careful study, a nut graf can be just as artful as the rest of your narrative. The nut is not just a kernel of knowledge, says David Robson, a features editor at *New Scientist*. It's a keystone. "You want to give a gist of the big idea behind the story, or at least the relevance of what you're reporting and how it will change the reader's life or understanding of the world," Robson says. Although you don't want spoilers to keep people from reading, you have to convince a reader to finish the story.

Choosing material to hold the story together can be one of the most difficult parts of the writing process, and it's different for every piece. "It's so easy to lose sight of what makes the story relevant and interesting," Robson says. "The nut graf is probably the bit that suffers most." To help him focus, he jots down "key attractions" while he's reporting so he can include them in the nut graf, which he drafts early in the writing process. When he struggles to figure out what those key attractions are, he imagines grasping for his friends' attention down at the pub. "What one detail would I drop into the conversation to stop their eyes glazing over and get them to ask me more questions?" he asks.

Robson, like many writers, uses the nut as a story's foundation, referring back to it frequently when drafting. Others seem to have nut allergies. Sometimes writers simply forget to include one or aren't sure what to include or why a nut might help readers. Others fear a nut graf might give away their punch line. Jocelyn Zuckerman, a contributing editor at *OnEarth* (and author of "Plowed Under," excerpted below) once butted heads with editors there over a food column she used to write. "I was frustrated," she says. "I thought it needed something to make it more conversational. This is a different animal than a news story, so I didn't think I needed a nut graf."

Love them or hate them, nut grafs are something your editor is likely to require, and there's a reason for that. Pooh Shapiro, health and science editor at the *Washington Post*, says she sees, in equal measure, excellent nut grafs and those that need finessing. "When someone needs help I tell them that our stories have to explain to readers: Why are we telling you about this particular thing now? Why should they care about it and make the time to read this story?" Shapiro says.

It's no accident that nut grafs are also called billboards. After days, weeks, or months reporting, writers often get so close to their stories that all the details seem clear. Readers have a different perspective. It's easy to get lost in the twists and turns of a complicated story—or even a not-that-complicated story. Nut grafs, or billboards, give readers a chance to preview a story's main themes and arguments from a distance before diving into the details. Think of nut grafs like the best billboards you've seen—a glimpse of the goodies waiting around the bend. That glimpse can intrigue readers to follow you to the end of your story. The nut can also function like a movie trailer: After introducing the story's characters and main theses, you can hook readers by foreshadowing tensions, reversals, or other suspenseful moments in your story.

While nuts should appear early, they don't have to come right away after the story's lede. Environmental journalist John Platt, who writes the *Scientific American* blog *Extinction Countdown*, keeps a scene-setting structure in mind. For stories about endangered species, he says, "I want to get people to care a little bit, understand the implications of what we're talking about, or sometimes emotionally feel that context before I actually spell the nut out for them." He watches his articles' "stickiness"—how many minutes people spend reading—to gauge how effective he was at keeping readers' attention.

Writing an effective nut graf requires knowing when to keep your lens narrow and when to jump out to a larger context, says Sam Fromartz, editor-in-chief of the Food and Environment Reporting Network (FERN), a nonprofit news organization that partners with publications to publish long-form stories. "You're constantly adjusting that lens," he says, and referring details back to the issue in a nut graf. When editing Zuckerman's story, which was produced in cooperation with FERN, Fromartz helped work the nut to move from a scene in a local diner to the wide expanse of the northern plains.

Even so, he says, sometimes a nut is unnecessary, especially if it becomes a crutch and doesn't work with the story. "When it's a really strong narrative and the story's just driving you, that can be strong enough to hook the reader that it doesn't necessarily need to be spelled out," he says. Other times a single, well-placed quote that captures the story's key meaning or significance can serve, on its own, as the nut.

But for most features and long news stories, a nut graf is necessary— and in some cases, it can help make a good story great. The following examples run the gamut from punchy quote to universal statement, from the almost-spoiler to compelling readers to satiate their curiosity. These nut grafs all work, and the writers and editors behind them explain how they were conceived.

From "The Story in the Stones," by David Robson for *New Scientist*, a story about stone tools shaping the human mind as told through the work of a flintknapper named Bruce Bradley:

> My interests lie elsewhere. The stone tools on the table in front of me are not just useful, they tell the story of our journey from simple ape to thinking human. Previous attempts to trace the history of the mind have relied on speculation as much as hard evidence but, over the past three years, Bradley's Learning to be Human project has taken a more precise approach to looking inside the heads of the people who made these tools. Combining findings about stone-tool construction with neuroscience, psychology and archaeology, we can now estimate the origins of the distinctly human mental abilities, such as when we first began to order our thoughts and actions, when our visual imagination blossomed, when we started to think about the past and the future,

and when we first played make-believe. There are even hints about the emergence of our capacity for patience, shame and suspicion—and the nature of our ancestors' dreams.

Near the end of reporting for this story, David Robson shadowed the flintknapper Bradley. The source was a godsend, his work carrying the narrative throughout the whole article. "Sweeping evolutionary stories can be a pain to write, since they easily grow into baggy monsters without any kind of shape," Robson says. "After all, evolution is a messy process with no clear direction or turning points, and you don't want to be too clumsy in molding that into a narrative."

After settling on Bradley's stone tools to shape the story, writing the nut came fairly easily. Robson intentionally chose first-person for both the lede and nut: "I dislike feature articles with 'disembodied' intros that don't gel with the nut graf. So I tried to write the transition as if my attention was wandering around the room until it finally settles on the tools, which seemed a more elegant way of segueing into the thrust of the feature."

This paragraph is long as nut grafs go—157 words—but the cadence heightens the experience of reading it. "You could imagine writing it more concisely; something like, 'We can now understand the forces driving the evolution of our intelligence,'" Robson says. "But it just felt a lot grander when I spelled it out in a list of experiences that we can all directly relate to."

In addition to hooking readers into the story and giving a bird's-eye view of the story, Robson aimed to emphasize the news angle. "The fact that they can read the thoughts of our ancestors in that level of detail felt far fresher and more exciting," he said.

From "Plowed Under" by Jocelyn Zuckerman in the left-leaning political magazine *The American Prospect*. The story, about land conversion in the Midwest, begins with pheasant hunters grumbling about disappearing prey. Then we get to the nut graf:

While few seem to be aware of it, a massive shift is under way in the northern plains, with ramifications for the quality of our water and food, and, more fundamentally, the long-term viability of our farms.

A study published in February 2013 in the *Proceedings of the National Academy of Sciences* found that between 2006 and 2011, farmers in the Dakotas, Minnesota, Nebraska, and Iowa—the Western Corn Belt— had plowed up 1.3 million acres of native grassland in order to plant corn and soybeans. "People had been talking about the land conversion," says Chris Wright, an assistant research professor at South Dakota State University and a co-author of the report, "but there weren't any recent numbers."

In contrast to Robson, Zuckerman followed a less-is-more philosophy with this nut graf. In fact, in the first draft of this story, there was not a nut to be found. The preceding paragraph opens with, "The region's game birds are in serious trouble." Zuckerman admits that at first she thought that was the nut. But the pheasants were only one part of a wider narrative.

Her editor, Fromartz, says on the first read, he kept coming back to the lede and asking, "Are the readers of *American Prospect* going to care about birds?" He returned the draft to Zuckerman, advising her to widen her lens from the birds, and to do it early. "The first draft had a lot of puzzle pieces, but there wasn't a lot of connective tissue," he says. "The nut graf is your chance to hook the reader—why these little birds in a place you've never been in your life should matter."

While this nut foreshadows a theme of small-farm vitality, much of the article focuses on policies, crop insurance, and pollinators on the plains. There's no hint of those topics in the nut—and by design, says Fromartz. "Once you get into policy, you can just feel your readers leaving," he says. "What was smart about this is she made the point that this is the greatest loss of these grasslands since the 1920s [in the next graf], and readers ask, 'What's up with that?'"

Zuckerman agrees. "I'm more interested in saying, 'Here's the problem,' and then later look at *how* is this happening?" she says.

From "Can What You Eat Affect Your Mental Health?" by Gisela Telis for the *Washington Post*, which opens with a woman's depression vanishing after she changed her diet:

> Corbitt had stumbled into an area that scientists have recently begun to investigate: whether food can have as powerful an impact on the mind as it does on the body.

Research exploring the link between diet and mental health "is a very new field; the first papers only came out a few years ago," said Michael Berk, a professor of psychiatry at the Deakin University School of Medicine in Australia. "But the results are unusually consistent, and they show a link between diet quality and mental health."

Using the one-two punch of an anecdotal lede plus scientist-quote nut let Telis set readers up for a compelling double storyline: following scientists committed to finding answers for their patients and women taking the initiative to change their lives.

Telis found an organic rhythm for telling the story after discovering her lede character early on. "I wanted to back out and get into the science of what her anecdote means—if anything," she says. The nut, she says, came easily once her interviews were in place.

The story could have been about an ignored field of science, or one that people are resistant to accept, Telis says. Yet after hearing the same explanation from multiple sources that "nobody's studying this, but we're starting to," Telis felt confident enough to make it the nut and let the rest of the story follow. She looked back in her notes to find the most striking and accessible quote. "I thought, how do I turn this personal story into the general?" she says. "And I try to do that as quickly as possible to move readers along."

From "The Last Meadows?" by Roberta Kwok in *Aeon*, a story about trees' "hostile invasion" of mountain meadows:

Trees are already on the move. Global warming has allowed forests to infiltrate meadows that were previously too cold or snowy, and grazing animals and fires no longer hold seedlings in check. A 2012 study led by forest ecologist Harold Zald at Oregon State University in Corvallis showed that trees have expanded from 8 per cent to 35 per cent of the meadow area in part of Oregon's Central Cascade Mountains over the past six decades. At two ridges in Canada's Rocky Mountains, the average meadow size shrank by 78 per cent from the 1950s to 1990s. Pines and larches are creeping into meadows in the European Alps too, and a 2009 meta-analysis led by the biologist Melanie Harsch, then at Lincoln University in New Zealand suggested that treelines have advanced to higher altitudes or latitudes at about half of surveyed sites worldwide.

This is not your traditional nut graf, just as Kwok intended. While Telis widened her lens, Kwok flashed hers around to different locations.

For this piece, Kwok knew that most readers see trees as victims of climate change, disease and other threats. To show trees threatening other ecosystems, she says, "I thought it might take a bit of convincing to get readers on board with that idea, so I wanted to back up that statement with data early on. I wanted to show readers that this was a real phenomenon and had been studied pretty extensively at sites around the world."

Like many of us, Kwok tends to write nut grafs that provide "mini-outlines" for her stories. But after a while, she says, that approach feels formulaic. "If there's another way that you can introduce the story and keep the reader interested, go for it."

At the end of the day, writing nut grafs might still feel like eating your veggies. But that doesn't mean you can't dress them up to be delicious.

28 How to Find and Use Quotes in Science Stories

Abdullahi Tsanni

WHEN ALEXANDRA WITZE was writing a story in July 2021 on the controversial question of whether to rename NASA's James Webb Space Telescope, she spoke to dozens of sources who had lots of opinions on the issue. To make sure that this diversity of opinion was captured, she carefully selected quotes from numerous sources. She interviewed three astronomers—Peter Gao, Saurabh Jha, and Johanna Teske—who all observed that, in their opinion, the telescope should be renamed.

Witze, a journalist based in Boulder, Colorado (who is also on *The Open Notebook*'s board of directors), mentioned all three astronomers in her story but used a quote from only one of the three, Peter Gao. She chose to do this because, while the other two interviewees made similar comments to Gao that they thought the telescope should be renamed, they also offered additional details, such as potential alternative names for the telescope. So Witze paraphrased their quotes in order to enrich the story without making it into a long series of quotes. "I didn't want people all saying the exact same thing," says Witze. "I had to work really hard to avoid that."

After conducting interviews, journalists typically have a lot of information and quotes to choose from. Deciding which quotes to use is a key step in the writing process, and one that requires skill and careful thought. Quotes serve many different functions in science stories. They can present a source's opinion, give factual information or context about how a study was conducted or about its results, suggest a metaphor or analogy to explain a concept or process, or express why a scientific discovery matters.

The number, length, and type of quotes you use will vary depending on whether you are working on a news or feature story. But one thing is constant: Quotes should always serve a purpose, and that purpose is to reinforce the main idea of your story.

Quotes Express Emotions

Science is a human process, performed by people. Yet scientific papers tend to be written in the passive voice and seldom give scientists the chance to express emotion or personality. Journalism is different. Science writers can use quotes to humanize the scientific method and bring the struggle, hard work, and joy of being a researcher to life.

"It's the one time we hear [scientists] speaking in their own words," Witze says. "The purpose [of researcher quotes] is to enrich the story by letting the voice of those close to the action be heard."

In a 2021 *Guardian* story by Oliver Franklin-Wallis about the development of the Oxford/AstraZeneca COVID-19 vaccine, this quote from Cath Green, head of a clinical biomanufacturing facility, captures the thrill of when her team managed to grow small amounts of coronavirus DNA into enough material to make trial vaccines:

"There's a stage where you have to centrifuge the material through a gradient, and all of the vaccine lines up in one layer. You can see it. Kathy, one of my production team, came out and said: 'Look at these babies!' We were like, 'Oh my God!' That was the day when we knew we had enough, and we could get it into somebody's arm in three weeks' time. That was great. I will remember that always."

Quotes offer a window into people's interior emotional lives—their feelings, reactions, and physical sensations—says Brooklyn-based science journalist Shayla Love, a senior staff writer at *Vice News*. This includes researchers, as the example above shows, as well as the people their research could affect.

In health stories, quotes from people living with a particular condition or disease help convey their subjective experience. "I find it best to hear straight from them," says Love. "Rather than summarize what they are going through or thinking about, I let them speak for themselves."

In a 2020 story on pedophilia, for example, she quoted a man going by the pseudonym Joseph Parker who described how his condition affected his sleep:

"As soon as I tried to release myself from wakefulness, my mind would sink into the pool of sexual energy, and I would feel this horrible sense of joy and happiness towards children."

This quote, as disturbing as it is to read, does express the person's feelings in a way that a summary made by a journalist could not. Quotes provide access to the lived experiences of people which nobody else would have access to, Love says, unless they also happen to be experiencing the same thing themselves.

Quotes Provide Details or Context

Reporters can use quotes to give context around why a study is important, why it is being conducted in a particular way, or what the results mean for the wider world. For example, in a *STAT* story by Megan Molteni, Yadong Huang, the coauthor of a study on preventing Alzheimer's disease, said:

"Developing new drug targets for Alzheimer's disease takes a lot of time and money, so we wanted to find a faster way to move drugs to patients."

The scientist elaborates further in another quote:

"There are many cellular and molecular changes in Alzheimer's disease patients besides plaques, but we usually don't talk about them," said Huang. "These results suggest that in order to treat Alzheimer's we should probably not target only one or two but multiple genes and multiple pathways involved in the disease."

Quotes can also provide the "bigger picture" as to why a scientific study is significant. For example, in a *Nature* story about NASA's Perseverance rover mission on Mars, Witze quoted planetary geologist Vivian Sun of the Jet Propulsion Laboratory in Pasadena, California. The quote encapsulates the immense ambition of that mission:

"Perseverance will be the first rover to seek the answer to the outstanding questions that we haven't been able to address with other rovers—was there life on Mars, and can we find evidence that there used to be life?" says Sun.

Quotes Present Opinions

Research is rarely black-and-white. Researchers have opinions and views on topics that science journalism can shine a light on, helping readers understand science as a human process, subject to debate and interrogation.

In a recent story about the historic approval of the first malaria vaccine, *Nature* reporter Amy Maxmen quoted a scientist named Badara Cisse, who expressed skepticism about whether public support for the vaccine would be sufficient, given its level of effectiveness.

> "I respect the researchers involved with this massive effort, but the reality is that so much money has been poured into this vaccine, even when the results from studies are disappointing.... I don't think a 30% effective vaccine would be acceptable for Americans."

In another *Nature* piece, Gayathri Vaidyanathan quoted two scientists who expressed their personal thoughts about India's ambitious carbon-zero pledge made at the COP26 climate meeting in Glasgow:

> "It's an ambitious target," says Apurba Mitra. . . . "It has put net-zero on the table."
> "It's great; a very bold announcement," adds Vaibhav Chaturvedi.

How to Select Great Quotes

Learning how to recognize which parts of an interview could produce great quotes can help make the writing process more efficient. Here are some qualities of a great quote to keep in mind: They contain critical information, vivid language, provocative ideas, strong opinion, or a unique perspective. "The ideal quote is one that summarizes, in clear language, a significant point of the story," says Witze.

While conducting interviews, New Delhi-based science journalist Sonali Prasad looks out for any strong expressions of emotion. "You can see the person either crestfallen, or excited, or moved by the conversation," she says. When Prasad observes such emotions, she makes a note of the timestamp on her recorder so she can later go back to her transcript to scan for worthwhile quotes. "This brings color to your piece," she says.

If a source gives her an answer that is unclear during an interview, Prasad rephrases her question in the hope of getting a more coherent response that she can later use direct quotes from. She also asks the interviewee to slow down, if need be.

Remember, you don't have to quote every source that you spoke to. To choose the best person to include, Witze asks herself a couple of questions: Is it a person whose voice has not been heard until now? Is it the person who has the most knowledge about the topic?

Witze also stresses that you should select quotes that accurately reflect the entirety of an interview. Researchers can veer off in interviews, perhaps talking about things they don't have expertise in, or revealing something they really shouldn't, she says. Witze is always careful, for example, when quoting early-career researchers who might be speaking too informally during an interview. "It's really easy for graduate students to mouth off to me about funding agencies," Witze says. However, this does not imply that journalists should quote only sources saying things that will make them look good. Nor is it the responsibility of a journalist to protect a source's reputation while they're being interviewed.

The number and length of quotes you use will depend on the type and scope of the story. For a feature story, says Akshat Rathi, a London-based reporter for *Bloomberg News*, you may speak to many sources and quote several of them. For news, you will end up quoting fewer people. But you don't need to treat quotes differently for news and feature stories. "A quote is like icing on the cake—it makes the story more interesting because of people's voice and expertise," says Rathi.

When to Paraphrase

Not all direct quotes will fit seamlessly into your piece. Some quotes may be very wordy, long, or contain too much technical language. They might be grammatically incorrect or clumsily phrased. Or they might be just plain boring or generic. If this is the case, one option is to paraphrase—putting a quote into different words without altering the meaning, and still attributing the idea to your source.

Rathi says he often paraphrases sources if quoting them directly might stop or slow the reader. But typically, when he does so, he includes a direct quote after the paraphrased material. "Keeping direct quotes also shows that you've had a chance to speak to the source," he says.

Paraphrasing is also helpful for breaking down complex scientific concepts, allowing you to add explanations or examples where needed. In those cases, says Abigail Beall, features editor at *New Scientist*, "It's always good to go back and ask, 'Is this correct?'"

One situation in which paraphrasing can really help is when sources are talking about numbers, data, or statistics. To simplify your story and stop the pace from dropping, you can take the nitty-gritty detail of the numbers out of direct quotes, but then include a quote from your interviewee discussing the significance of the numbers or reacting to them.

There are some times when it is preferable to avoid paraphrasing—for example, when reporting on legal issues. Using direct quotes ensures that you are accurately reflecting what the source told you. "It offers you safety in a way, as a journalist, because nobody can suggest that you've changed what somebody has said," says Fiona Broom, features and investigations editor at *SciDev.Net*.

When and How Much to Edit Quotes

There are many situations in which you might wish you could tweak a quote—sometimes, sources just don't express themselves as clearly, eloquently, or succinctly as we might want them to. But most media organizations strictly forbid changing quotes, for the simple reason that the act of putting words within quotation marks is, in effect, a promise to readers that the quoted words actually came out of a person's mouth. Some media organizations make exceptions for making minor grammatical corrections; others, such as the Associated Press, make no such exceptions. Before you change anything about a quote, always check in with your editor, advises Witze, because publications have different policies on whether and how much quotes may be changed. "If I turn in a story and the editor thinks the quote needs to be tweaked a tiny bit in order to be more readable, and if the publication permits that sort of change, then that is the only situation in which I would ever change [a quote]," she says.

When confronted with quotes that aren't grammatically correct, it can be tempting to make small fixes to quotes. But that may set a dangerous precedent. Instead, editors at *Undark* simply paraphrase, rather than rework anything within the quotation marks, says Brooke Borel, an editor there. She adds that it's best practice not to remove words within quotation marks, nor to put sentences together that weren't spoken together. "There are ways to paraphrase or otherwise indicate to the reader what the speaker was trying to get across," Borel explained in an email.

Likewise, Prasad says she avoids tinkering with quotes. "Sometimes, a simple rearrangement of sentences or words can alter meaning or context," she says.

When using quotes from an interview that was carried out in a different language than that of your article, it's important to work with a professional translator, and to consider also using a fact-checker who works in that language. Prasad did both while making a documentary on the west coast of India, when she interviewed fishermen in a local language that she didn't speak.

For representing different dialects, there seems to be no satisfactory guidance. On one hand, you want to make sure that you aren't unintentionally creating a caricature of a source by writing their quotes in a particular way. On the other, you want sources to sound like themselves, even in print. "I think there are arguments for both preserving and not preserving dialect that make sense, and that editors and journalists have to approach this with sensitivity," says Borel.

As crucial as quotes are in science stories, distilling great ones from a mess of interview notes can be overwhelming, and it requires skills. Always keep three things in mind: clarity of the quote (is it short and understandable?), content of the quote (does it illustrate a major point that needs to be made in the story?), and the person providing the quote (are they speaking to their area of expertise and are they the correct person to be making this point in the story?). When used with sensitivity and purpose, quotes are a powerful tool. "It's not enough to use a quote just because it sounds good or the source said something nice," says Witze. "Don't just throw in a quote if it doesn't serve the purpose of your story."

29 Like Being There: How Science Writers Use Sensory Detail

Jyoti Madhusoodanan

At this time of year, with new growth laying a haze of green over the wet fields, the farm country around this small town smells faintly but distinctly of manure. It's a rich, warm aroma, appropriate to the place that bills itself on road signs as "Canada's foremost cattle county." But follow the dip in Hwy. 4 over the Saugeen River and down into Walkerton, and the smell changes noticeably. It is acrid and ammoniac. It hits you in the back of the nose, and it is weirdly familiar. It smells like a swimming pool. It's bleach. All the people living in Walkerton, and most of the objects out in public—doorknobs, store counters, cafe tables—have been washed or swabbed with a potent mixture of chlorine bleach and water, the most effective way to kill the bacterium that has contaminated their water system and invaded their lives.

—from "Ontario Town Battling Bacteria, Fear," by Maryn McKenna for the
  Atlanta Journal Constitution, May 30, 2000

WHEN MARYN MCKENNA began writing a series of newspaper stories on a Canadian *E. coli* epidemic for the *Atlanta Journal Constitution* more than a decade ago, she was surprised by the number of readers who connected with this small Ontario town through her articles and who wrote her to say, "You made me feel like I was there."

They didn't relate to the feeling of living in a town where the local aquifer's fluid dynamics were particularly unique, or to the reasons for the epidemic. But everyone knew what bleach smelled like.

Memorable science stories captivate not only because they hold our intellectual attention, but also because they grab us by the senses, weaving smell, touch, and taste through abstract concepts like the epidemiology of bacterial infections. At a session at the Santa Fe Science Writing Workshop in May, 2012, Maryn McKenna shared tips for gathering and including such details to make for more vivid science reporting. To explain the need for such detail, she drew a parallel between science reporting and semanticist (and later congressman) S. I. Hayakawa's ladder of abstraction, explaining why these concepts are so powerful and why it is so important for a science writer to think about using them.

In the 1930s, Hayakawa proposed that we process experience with the help of what he called the ladder of abstraction. He drew it as a literal ladder, with very concrete details at the bottom and very abstract ones at the top:

- At the top is God, and at the bottom are worn rosary beads.
- At the top is War, and at the bottom is blood on a doorstep.

As science writers, we spend a little time at the top of the ladder—evolution!—and a whole lot of time in the middle, which is where concepts and data and explanations live. "What sensory detail does is to drop us like a plumb line to the bottom of the ladder, down to the place where people's attention is engaged. Once we do that we can bring them back up the ladder again," McKenna said. "Rarely does science writing get right down to these bottom levels and engage a reader's senses and emotions."

Stories that incorporate these bottom-rung sensory details transport us—whether it is into towns being decimated by an epidemic, into dark caves filled with dying bats, into a Canadian distillery town beset by a mysterious fungus, or into the world of an octopus, where intellect is inextricably linked to touch and taste.

How do writers report and use sensory detail to create vivid stories?

"Notice what you notice, and write it down," McKenna says. "If it's not in the notebook, it's not in the story. Capture every kind of detail—the more the better."

Don't wait to write down the details that impress you the most, adds Michelle Nijhuis, whose 2011 *Smithsonian* story "Crisis in the Caves" won an Outstanding Article award from the American Society of Journalists and Authors. "I've been in many situations where I'm having such a vivid experience that I'm sure I'll never forget it," she says. "But then sure enough, the next experience is just as vivid and I forget the first one. Even if you think you will remember something forever, write it down." As an alternative, she also suggests talking to oneself. "I know some people who will talk to themselves into a recorder at night in their tents while in the field," she says.

McKenna also recommends a phone camera to take quick snapshots and capture visual detail to include in words later. *Wired* editor Adam

Rogers, who won the AAAS Kavli Science Journalism award for his 2011 story "The Angel's Share," about a mystery fungus in a distillery town, also takes pictures while reporting to add scenes to a story. But, he suggests, "Smell and taste can be captured better by writing than by any other media. A camera can't capture chemical sense detail the way writing can, and I think it's more efficient to capture a scene with smell and taste rather than something visual. Stories almost always have art and pretty pictures to go with them, but we rely on the text to convey these chemical senses."

Author and broadcast journalist Sy Montgomery has reported stories from cloud forests and a canoe on the Amazon River, and still relies on her notebook more than anything else. "I will use a recorder in the field, but my notebook travels everywhere with me," Montgomery says. "Also, I had an orangutan eat my tapes once, and I was really glad I had my notes then. Nothing has tried to eat my notebook."

Once details are written down, photographed, or otherwise recorded, how do these writers decide which nuggets to include and which to discard? Not every detail needs to be included with every story. For her 2011 *Orion* story "Deep Intellect," on the octopus's intellect, Montgomery spent several days in an aquarium, where everything either "smells like a fish or is a fish." But since her story was about chemosensory receptors for touch and taste, the smell of an octopus seemed irrelevant. "On the other hand," she adds, "it was important to me to convey to readers the precise temperature that the octopus lives in. My hands were too cold to write when I took them out of the water, but the octopus senses everything at this temperature."

McKenna advises that writers only use details that add to their stories and characters: "Angle them towards the subject to make them count." Rogers concurs that not all details work, citing the all-too-ubiquitous researcher profiles that describe a professor wearing a "professorial sweater" in his office at the end of a hallway in a university building. "This tells me nothing about why I should care about this person. Spare me these cheap details," Rogers says. Along with avoiding the obvious, Rogers also advocates staying away from far-fetched analogies that a reader can't necessarily relate to. "Don't tell me something felt like hang-gliding," he says.

Instead, Rogers suggests using metaphors "to add a pop at the end of a technical explanation, rather than as a crutch to prop your scientific explanation up." In "The Angel's Share," he describes scientists shaking their heads over a misclassified fungus "like a plumber shaking his head over a homeowner's attempt to patch a leaky pipe."

Nijhuis sometimes seeds metaphors in the field to see if they take hold with her sources. "Sometimes, I'll just make a statement out loud—like 'this animal feels like a fancy sweater'—and the people I'm with build off that and give me more material to work with," she says. To describe the feel of an octopus's suckers on her hands, Montgomery turned to metaphor, likening the experience to "an alien's kiss—at once a probe and a caress." Octopuses are slimy, she explains, but she wanted to stay away from words typically associated with disgust. "I wanted to convey the delight that I felt to be touching an octopus, and it was important for me to be in the narrative to clarify that these are my perceptions," she says.

Like many sensory experiences, delight in an octopus's touch is a subjective emotion, largely dependent on the writer. Montgomery resolves this by using a first-person narrative. Nijhuis emphasizes that recording personal experience is critical, even when the writer is not in the story as directly. "When I was standing in the cave, I was observing my own reactions as well," she says. "Did I feel cold? Was it damp? Even on the phone, your impressions of the person at the other end are important to the angle of your narrative."

Rogers adds that smell, touch, and all other sensory details are subtle ways in which writers put themselves in the story, but posits that this is a contract between the writer and the reader. He says, "To me, it is like promising the reader I will take him into this journey, like telling them that we're in this together, and watching what unfolds as the story progresses. This is why we still send reporters out into the field instead of doing everything on the phone or email, because at the end of it, we still want to feel what a person experienced while they were there."

30 Eradicating Ableist Language Yields More Accurate and More Humane Journalism

Marion Renault

AS DISABILITY HISTORIAN Aparna Nair and I discussed how regularly disability is portrayed as a pitiful tragedy or a personal failure, exasperation washed over us.

"It is maddening," I said.

"It's entirely maddening," she replied.

Then my mind lurched, tripping over the word "maddening." Its roots trace back to the Old English for "out of one's mind," the Proto-Germanic for "changed for the worse," the Old Saxon for "foolish," the Gothic for "crippled, wounded." Seven weeks of reporting this story had finely sharpened my awareness of ableism—and yet the word left my mouth without the slightest friction. Nair and I agreed: My slip-up exemplified the insidiousness of ableist language.

Ableism evaluates the worth of bodies and minds based on socially constructed criteria of normalcy, performance, and intelligence. Those deemed "deviant, abnormal, defective, subhuman, less than," as disability activist Lydia X. Z. Brown describes, are erased, marginalized, or abandoned—often with the linguistic assistance of slurs or more subtle pejoratives. Ableism inverts the simple fact that disability is normal, an intrinsic and ubiquitous feature of the human experience.

A "disability," as defined by the Americans with Disabilities Act, is "a physical or mental impairment that substantially limits one or more major life activities." The term includes chronic and terminal illnesses; communication and neurological conditions; and developmental, hearing, intellectual, learning, psychiatric, physical, sensory, and vision disabilities. Disability can be visible or invisible, temporary or permanent, life-long or acquired. Altogether, disabled people form the nation's largest minority group, one we're all likely to belong to at some point in our lives. "No other social identity category is so porous and unstable," writes bioethicist and disability-justice scholar Rosemarie Garland-Thomson in the essay collection *About Us*.

Journalists have helped propagate ableism by using language that casts disability—directly and indirectly—as abnormal and socially unacceptable. Some portrayals of disabled people are undeniably negative, painting them as incapable or criminal or miserable. Others "elevate" disabled people by characterizing them as saintlike, magical, or inspirational. In either case, disabled bodies and minds are set apart.

In reality, says Emily Ladau, author of *Demystifying Disability: What to Know, What to Say, and How to Be an Ally*, disability is "not a bad thing. It's just a natural part of the human experience." Like every human's existence, the disabled experience encompasses suffering and pleasure, joy and sorrow, content and discontent; it varies wildly from person to person, from day to day; it is vast and contradictory, complex and ordinary. Words that reduce disability to a single quality commit harm by failing to capture the texture, fluidity, and three-dimensionality that any person—disabled or not—contains.

For the flattened, unnuanced, and negative societal view of disability to change, journalists must accept their role in narrating disability, says Aminata Sanou, a journalist with *Burkina 24* in Bobo-Dioulasso, Burkina Faso and a trainer in gender and inclusion. "We are society's mirror." If journalists don't shift their portrayals of disability, the general public's view of people living with disabilities won't budge either, she says. "If we decide to remain in a habitual state, it's clear nothing will change." Ableism is a chain, Sanou says, and "we shouldn't be ashamed to break it."

Start, but Don't Stop, with Glossaries and Style Guides

Journalists can begin by familiarizing themselves with freely available glossaries and style guides for covering disability. These valuable resources offer clear guidance on outdated and offensive ableist terms such as "crippled," "lame," or "spazz."

But you shouldn't plan to memorize some "no-no" words and call it a day. All language is fluid and evolving; today's standards are not tomorrow's. And as my "maddening" mistake exemplifies, ableist language is so deeply woven into our speech that perpetual vigilance is essential. "It's something we all have to actively keep in mind," says Rosemary

McDonnell-Horita, an independent disability consultant in California who has a mobility disability and uses a wheelchair. "You're not just going to wake up tomorrow and say, 'I'm stripping all ableist language from my vocabulary.'"

So journalists must ground these efforts in a genuine understanding of ableism. Ableism maintains its power through the continuous dehumanization of disabled people, and language is a key tool for accomplishing this. The need to avoid ableist language is thus not about avoiding hurt feelings. Instead, it's about confronting stakes that are grave, violent, oppressive, and ongoing: forced institutionalization and incarceration; social isolation; unmet basic needs; and barriers to entering schools, workplaces, and public spaces.

When teaching disability history, Nair says, many of her students are shocked to discover these horrors have survived into the present. "There's a tendency to want to believe that the violence is over . . . and we are not capable of that anymore," she says. "Nothing could be further from the truth."

For example, one in four disabled adults in the United States lives under the poverty line, compared to fewer than one in ten nondisabled Americans. In 2022, only one in five people with disabilities in the US was employed, compared to two-thirds of nondisabled people. At the same time, it's still legal to pay disabled people subminimum wage. Disabled people are almost four times as likely to be the victims of violent crime; disabled children are also nearly four times as likely to experience physical, sexual, or other violence. Around the world, states fail to deliver on promises to support or protect disabled people.

The historic and continued neglect and abuse of disabled people remain encapsulated in many commonly used words. When you look into their histories, says s.e. smith, an independent journalist who helped launch the resource Disabled Writers, "sometimes you discover really ugly things."

Early-20th-century eugenicists used terms like "imbecile," "moron," "idiot," and other terms to connote degeneracy, defectiveness, and unfitness. Their dehumanizing effect paved the way for involuntary sterilization and segregation into horrifically abusive and neglectful institutions. That violence persists today: During the height of the COVID-19 pandemic, for example, intellectually disabled people were a lower priority for lifesaving treatments in some states.

The term "midget," a slur against people with dwarfism, comes from "midge," and was co-opted by a carnival and circus industry that profited from exhibiting people with physical disabilities like zoo animals. "You are calling a human a fly," says Cara Reedy, a journalist, director of the Disabled Journalists Association, and a dwarf. "That's violent."

Many people casually lob terms referring to mental health conditions as throwaway insults—"crazy," "insane," "bipolar," "OCD," "schizophrenic," "psychotic," "catatonic," "demented," "loony," "mad," "nuts," "deranged," "bonkers"—even though such accusations once justified locking someone up in an asylum indefinitely, and even now, having a mental health diagnosis can lead to poorer access to health care, employment discrimination, and loss of legal autonomy.

"Lame"—a term whose meaning as "maimed" or "weak" dates back between 1,500 and 400 years across several archaic languages—came to mean "uncool" in the twentieth century. But in antiquity, children with physical disabilities were socially scorned, mutilated to boost their value as beggars, and even murdered by parents who abandoned them to die in the woods with feet bound. "Let there be a law that no deformed child shall live," wrote Aristotle. Today, the perception that people who need physical-access accommodations have lesser value undermine inclinations to build ramps or elevators.

Despite these deep-seated cruelties, ableist terms remain, paradoxically, both colloquial and (sometimes) hard-to-recognize. "It doesn't even come up on people's radar," McDonnell-Horita says, even though it's baked into our everyday vernacular. But regardless of whether they register as such, Nair says, "these labels are inextricable from those histories."

Beware Figurative Language

"Disabled" is not a slur—it's a simple description of reality. But some people struggle to say the word out loud during sensitivity trainings, McDonnell-Horita says. "You can hear the hesitation in their voice." She and others credit that discomfort to the proliferation of euphemistic language—"differently abled," "diffability," "specially abled," "handi-capable," and "special-needs." "Really what you need to say and what you mean to say is 'disability,'" McDonnell-Horita says.

Other times, we use figurative language to more subtly tie disability to notions of dysfunction, malice, or deviance. That which is "crippled" or "paralyzed" is damaged. A message "falls on deaf ears" when ignored. A politician with no power is a "lame duck." Someone "turns a blind eye" through intentional ignorance. "Leaning on a crutch" means exces-sively relying on assistance. "We so often use disability in negative con-texts without even realizing it," says Ladau, who has genetic disability called Larsen syndrome that affects her joints and muscles.

Especially pertinent to journalists, many demeaning disability meta-phors doubly commit the writing sins of being inaccurate and clichéd. "I think of metaphors like that as lazy," says Amanda Morris, a disability re-porter for the *Washington Post*. A person can be ambushed regardless of whether they have a "blind side" or visual disability. A speech did not lit-erally "fall on deaf ears" unless the audience was a group of people who are exclusively deaf and/or hard of hearing. Terms like "special needs" or "differently abled" deny the fact that everyone—*everyone*—has unique needs and abilities. A person is not literally "bound by" or "confined to" a wheelchair. "My wheelchair is my freedom," McDonnell-Horita says.

Medicalized language—"disorder," "impairment," "deformity," "ab-normality," "defect"—frames disability as a pathology to be cured. ("Con-dition" is usually, but not always, a more neutral word.) Characterizing someone as "suffering from," or being "stricken by," or "afflicted with" their disability imposes an emotional assumption about life quality that may not reflect an individual's lived reality.

Nondisabled people may struggle to notice abstract ableist language and take it seriously. "A lot of people feel like, 'It's not that deep,'" says Wendy Lu, a senior staff editor at the *New York Times* who is multiply dis-abled. "But when you are disabled and you constantly hear these idioms every day it really makes it seem like we're just not there."

Disabled people are forced to navigate a world that often considers them subhuman. Brushing that reality off or bemoaning efforts to ad-dress it as hypersensitive word policing is itself ableist. "It stings to see your existence used as a shorthand for bad," says smith, who challenges fellow journalists to ask themselves: "Why do you feel like, when you're describing something that is bad or negative or unwanted, that disability has to be your go-to?"

Be Ready to Face the Gray

Given the disabled community's diversity, it makes all the sense in the world that there is no universally agreed upon way to talk about disability. "We're not very dictatorial in our guide," says Kristin Gilger, executive director of the National Center on Disability and Journalism (NCDJ), which publishes a style guide for journalists. "There are some things you say, 'Just don't do this.' But in many cases, it depends." For example, as the guide notes, many people consider that using the term "severe" to describe a condition "implies judgment" and advocate for using the word "significant" instead. But not everyone agrees with that distinction. Many people with myalgic encephalomyelitis, for instance, embrace the terms "severe" and "very severe"—only lamenting that they're not strong enough to evoke the most horrific and extreme versions of an illness that has historically been dismissed. That variability in opinion underscores why journalists writing on disability and chronic illness should, as the NCDJ guide suggests, "proceed with caution." The NCDJ resource and others like it, Gilger says, should be viewed not as simplistic how-to guides, but as tools to empower journalists to thoughtfully handle reporting about disability on their own.

Ambiguity also rules the sometimes-tricky terrain of what terminology to use to refer to disabled people. Some people (and particularly members of the autistic and Deaf communities) prefer identity-first language—"disabled person," "autistic person," and so on—which embraces disability as an inseparable part of one's personhood, just like one's race, class, gender, ethnicity, sexual orientation, and so on. Others prefer person-first language: "person with a disability." In the past, the NCDJ (and other research and advocacy organizations) encouraged journalists to default to person-first language; after pushback, they revoked that blanket advice. "Instead, we hope you will double down to find out how people would like to be described," the guide now reads. "We encourage you to confirm on a case-by-case basis."

When deciding whether to refer to someone as *being* ADHD or as *having* ADHD, as *being* deaf or Deaf or hard of hearing, as *being* autistic or *being* an autistic person or *having* autism, the resounding advice is to simply ask a person what they prefer and honor the answer. When not referring to a specific person—or when there is no clear consensus on what a particular disability community prefers—the advice is not as cut-and-dry. You might draw from communications or guidance from relevant, disability-led organizations or intermix person-first and identity-first language throughout a piece; you should still be careful not to lean on whatever language is used by "third parties" like parents or doctors.

Such questions are often complicated by other language considerations, too. For example, some members of certain illness communities note that the seemingly neutral term "patient" can reduce a person to an identity of illness and erase other aspects of their existence, and for that reason they prefer to avoid the term. So, for example, some might prefer "person living with HIV and AIDS" to "AIDS patient." (Again, it's best practice to ask an individual their preference.)

Journalists might enter other gray areas when they encounter seemingly unavoidable ableist language—science's "double-blind" studies or the quotidian "handicapped parking." These examples may seem to leave no room for options, but there are several possibilities for sensitively handling such language. We can choose to use these terms, put them in quotation marks, and explain why they're controversial. We can swap them out for more neutral language like "anonymous study" or "accessible parking," and explain any potentially unfamiliar terms to readers.

One potential exception, Lu says, is if a prominent source uses an ableist term or phrase in a critical news moment, in which case "you might have to include it." Alternatively, she says, you might paraphrase or choose a different quote. Consider a hypothetical quote in which a researcher describes an aspect of their work others might find surprising: "It's crazy, but it happens all the time." The writer could cut "it's crazy" from the direct quotation; they could also press the scientist some more in order to get a quote that didn't rely on ableist language to make its point.

In some cases, ableist language is enshrined in law or governance.

Several French-speaking countries offer health insurance or other forms of state support for "invalids" or "the handicapped." Similar to references to US government disability "benefits," public officials in Burkina Faso refer to disability "privileges" or "advantages," Sanou told me, rather than framing certain government programs as protecting their civil rights and providing adequate support. Another example is the many US agencies that still use the phrase "special education" to describe support for disabled students. And in 2016, despite pushback from disability advocates, several Indian government agencies were renamed, branding people with disabilities as "divyang," a term that means "possessed with divinity" and reinforces the stereotype of disabled people as magical or saintlike. That same year, in Japan, authorities refused to identify the victims of a terrorist who stabbed 19 residents of a disability care facility to death and injured 26 more; they defended the choice as a means of preventing the deceased's families from being discriminated against for being related to a disabled person. The decision ultimately made it impossible for journalists to cover each individual loss of life with human breadth and depth.

Disability also has distinctive cultural nuances that journalists must navigate thoughtfully. A majority of disabled people—80 percent—live in the Global South. Western standards when it comes to ableist language cannot simply be imposed, Nair says. "That's not productive if you really want to understand and communicate what it means to be living with a disability in the Global South." Instead, it's critical to respect and reflect the unique societal forces that shape an individual's experience of disability and keep in mind that language itself functions differently across contexts.

For example, when doing ethnographies of women with epilepsy in Indian hospitals, she avoids the word "epileptic"—even though that's how Nair, who has epilepsy, self-describes. Instead, she refers to them as they do: "sick." Similarly, Nair refers to people as having "Hansen's disease," not "leprosy," which offers a liberating distance from historic stigma in the more than 120 countries where the bacterial infection still persists. "I know the weighty past of that particular label," she says.

Discrimination against people with disabilities is sometimes rooted in theology, Sanou explains. In Burkina Faso, for example, many deeply religious people view disability as a divine punishment for their personal moral failures or those of their ancestors. "Public opinion doesn't necessarily have a compassionate view toward people with disabilities," she says. And so, in her reporting, she intentionally opts for the person-first "people living with disabilities" over the identity-first "disabled people." In Burkina Faso, she says, "it's less heavy to carry, I think."

From Anti-ableist Language to Anti-ableism

Journalists have an obligation not to just to watch our language, but to actively counter ableism in ourselves and our readers. "We live in an ableist society and journalists come from society," Reedy says. "Whether we're in a newsroom or not, we're all ableist."

Sanou, for example, focuses on individual disabled people in her stories, such as a man who had lived for 10 years as HIV-positive without telling anyone in his family. "The more we personalize, the more we humanize, the more we solicit compassion," she says, "the more we enter reality."

Journalists who want to explore more truthful and evocative ways of depicting disability can look to essay and poetry collections like *Beauty Is a Verb*, *About Us*, *Disability Visibility*, and *Body Language*, which offer a portal into the endlessly imaginative ways that disabled writers describe themselves in and on their own terms.

We can also reconsider professional norms, from interviewing sources to packaging stories. During a 2019 investigation into schools exiling disabled students to padded, cell-like "quiet rooms," for example, *ProPublica Illinois* journalists included children's drawings showing how it felt to be locked up for hours on end. In 2020, in a feature on pandemic isolation and dementia, the *Washington Post* included brief text conversations with a woman about her personal struggle: "I not talking with the whole sentence anymore," she texted. "Not got balance. Painful cramping." the *New York Times* made it possible to print its special coverage of the ADA's 30th anniversary in Braille. For the *Washington Post*, Morris, who is hard of hearing and a CODA (child of deaf adults), regularly appeals to her audiences through multiple senses with visual diagrams and audio clips.

Addressing language is only a starting place for addressing ableism as a whole. "How many racists don't say the n-word anymore?" Reedy asks. "You can stop saying the r-word, but you can still [fail to] talk to an intellectually disabled person when the story is about intellectually disabled people," she says. "You have devalued that person by not speaking to them." Reedy adds that "while the two are not the same, the function is the same, in that you can continue to be a racist without saying the n-word. You can also change your language around disability and still be an ableist." Eric Michael Garcia, a senior Washington correspondent for the *Independent* and author of *We're Not Broken: Changing the Autism Conversation*, agrees. "You could remove every 'wheelchair bound,' every 'falls on deaf ears,' and still wind up having very flat, very dry writing about disability," he says. "What you should be aspiring to is the wholeness of the disability experience."

31 Good Endings: How to Write a Kicker Your Editor and Your Readers Will Love

Robin Meadows

SOMETIMES I WRITE a story's ending first, and sometimes it pops into my head when I get there. Other times it feels like I've already said it all and I struggle with the kicker. But easy or hard, endings deserve as much care as beginnings. "While we obsess about beginnings, we often don't spend enough time sculpting our endings, or kickers, and that's too bad," Michelle Nijhuis wrote in *The Science Writers' Handbook*. "Endings are our last word to the reader, and often what readers remember most."

So how do you write an ending that sticks? Here are three editors on kickers they love and why they love them, and the writers on how they did it. Each kicker is preceded by a bit of context, including a short summary of the story.

Circle Back to the Beginning

"The Spy Who Loved Frogs"
By Brendan Borrell for *Nature*

BACKGROUND ABOUT THE STORY: To track the fate of threatened species, a young scientist called Rafe Brown must follow the jungle path of a herpetologist who led a secret double life. The story opens with Brown flipping through a 1922 monograph by herpetologist Edward Taylor and stopping in awe on a page documenting Taylor's discovery of a rare gecko in the Philippines.

KICKER: Back at the University of Kansas, Brown takes a seat inside an archival library and dips once more into some of Taylor's work, including the battered leather books that the man used for his field notes and specimen catalogues. Paging through one of those catalogues for the first time, Brown is stunned to find that Taylor had crossed out the name attached to an Asian spadefoot toad that he caught on Mindoro Island—a strange, gangly creature that crawls rather than hops. Next to it, Taylor had written, "new sp!!" As recently as 2009, Brown had designated it as a new species, *Leptobrachium mangyanorum*, because it was so different from previously described relatives.

"Ed was way ahead of us," says Brown. "Why he never named it, we'll never know. But it's pretty satisfying to come along 90 to 100 years later and arrive at the same conclusion."

THE EDITOR'S TAKE: Satisfying endings circle back to the opening scene, summarize the story's main points, point out future directions, and often include a pithy quote, says Brendan Maher. This ending does all that and more. "What sets this one apart for me is that it actually includes a moment of discovery," he says. "It reserves a new piece of information—usually a no-no for story craft. And in it, the past and present are meeting in real time. I really couldn't have asked for anything more apt."

THE WRITER'S TAKE: The ending is an opportunity to remind readers of the story's theme, give them closure, and leave them with something to think about, says freelancer Brendan Borrell. He knew this ending was "the one" from the first outline of his story. "I think the reason it works so well as an ending (and not a lede) is that it is this honest-to-God aha moment for Brown that echoes what we've learned before (Taylor was right!) but with a little bit of a twist," he says. "If you put it anywhere else in the story, it simply wouldn't pack as much of a punch."

Borrell also credits editor Brendan Maher with suggesting a new beginning that made this ending even better. Borrell originally led with a dramatic moment with Brown in the field, and Maher suggested leading with a scene where Brown pages through Taylor's monograph. Maher told Borrell, "I know it's not quite as action packed as what you have to start it now, but I really like how it ties the present generation and the past together. If we were to start with something like that, it would make the ending circle back really quite nicely to him once again reviewing a book touched by Taylor and kind of seeing the genius on the page."

Tie Everything Together

"Engineering New Organs Using Our Own Living Cells"
By Steve Volk for *Discover*

BACKGROUND ABOUT THE STORY: Inspired by the regenerative abilities of an amphibian, Anthony Atala is driven to save lives by rebuilding organs. The story opens like a fairy tale, with a "once upon a time" anecdote about a little boy with a failing bladder and a "wizard" (Atala) who miraculously grew him a new one in a petri dish. The story's ending evokes a couple of key scenes from the story. One describes a TED talk: "Atala held out a pink, newly printed kidney in his gloved hands. The sense of wonder, awe, even mystification, was evident in the crowd's feverish applause." Another scene relates an epiphany Atala has when he finds a rock on the beach that's shaped like a kidney and even has a line across it that looks like the line between two parts of the kidney: He realizes that he doesn't have to grow a whole new kidney—all he has to grow is a "wafer" of kidney tissue to insert at that line. The body will do the rest.

KICKER: Atala continues to work on creating whole new organs. But he also has a team working on the model that occurred to him on the beach: Harvest and grow some healthy cells from a patient's damaged kidneys. Concurrently, decellularize a pig kidney, leaving only the casing. Then repopulate the organ with the patient's cells. Insert a section of that new kidney tissue, equal in weight to maybe 20 percent of the existing organ. With no cells from the pig, the recipient's body should accept this new section of kidney.

Atala is also pursuing this "wafer" model of creating partial transplants for other organs.

Though Atala always remains circumspect about the status of his projects, he says this partial transplant model is different: That team is far along in the process, successfully placing kidney cartridges into animals for trials lasting several months. The major problems, he says, all appear to be solved. Relatively speaking, partial transplants are closer. The most practical solution may not be as dramatic, or garner as much publicity as creating a whole new organ. Yet millions of happy ever-afters beckon. Because he saw the answer when it washed in with the tide—the ocean rolling back in from the future, sounding like an echo of a mystified crowd's applause.

THE EDITOR'S TAKE: The best endings give a sense of resolution and of propulsion—ideas that carry on after the story concludes—without feeling forced, says Becky Lang. She particularly likes this ending because it circles back to the fairy-tale theme that runs throughout the profile, it alludes to scenes earlier in the story, and it is true to the extreme sense of caution that pervades Atala's thinking. "By coupling this wizardry with the real-world caution—and still having the pull of potential in there—Steve weaves all of it back together in one ending and sends readers off with their brains completely engaged," she says.

THE WRITER'S TAKE: Steve Volk looks for the ending and beginning as soon as he starts researching a story, and he often finds them at the same time. "The best endings echo the beginning in some essential but surprising way," he says. "So, often, realizing where a story should end immediately triggers a thought about where it should begin, and vice versa."

When Volk is ready to write, he outlines possible scenes to include and usually chooses those that evoke the greatest emotional response for his beginning and ending. That's what he did for this story, so he knew the end would revisit a scene where the shape of a rock on a beach inspires Atala to pursue partial organ transplants.

But Volk didn't know exactly what he would write until he got to the end. He drew his inspiration from another scene in the story. "I included an anecdote near the top . . . in which a demonstration of his printer technology causes the crowd to erupt in the kind of applause that is usually reserved for rock stars," he says. "And so there I was, thinking about the applause Atala has received and imagining him on the beach. I started imagining the setting itself. I thought of the feeling of sand under my feet and the constant pounding of ocean waves and—there it was. The roar of a crowd, the roar of the ocean."

Make It Personal

"Ecosystems 101: Hard Lessons from the Mighty Salmon Runs of Alaska's Bristol Bay"

By Ray Ring for *High Country News*

BACKGROUND ABOUT THE STORY: The world's longest ongoing salmon research reveals the astounding complexity of wild ecosystems. The story's kicker is a stand-alone ending that tells the story of the author's personal experience: the chance to swim with salmon (at the invitation of researcher Jonny Armstrong, whose work is included in the story.)

KICKER: About 10 years ago, on the bank of Washington's Snohomish River, surrounded by roads and farms and cities, I met a guy who put on a snorkeling mask and swam with the remnants of the salmon run there— just for the love of it. I put it on my bucket list: Someday, somewhere, before I die, swim with salmon.

One afternoon here, on Sam Creek, I got my chance. First, Armstrong donned his snorkeling gear and a dry suit and slipped into the cold, clear water to take close-up photos of the vivid red sockeye in a pool where the creek meets Lake Nerka. Hundreds of salmon hovered side-by-side, facing upstream, getting ready to make their run, and they tolerated Armstrong slowly easing into their formation.

Then we walked up the creek through a litter of salmon carcasses, past fresh bear and gull prints in the sand, finding smaller pools where more sockeye hovered. I wrestled into the rubber suit, adjusted the facemask, and lay down in a pool. Almost instantly, it seemed, I was surrounded by the big reds.

As they undulated to keep themselves in formation, some brushed against me with their bodies, and some swiped their tails against exposed portions of my face. A few even wriggled under me, one by one forcing their way between my chest and the sandy bottom.

Shifting slightly upstream, I turned to look at them head-on. Dozens faced me, just a few inches away, their jagged teeth exposed through gaping, elongated jaws—another physical change that occurs as they spawn, as if they're reverting to a completely primitive form that matches the landscape. Their gills flexed water in and out, extracting oxygen, and their eyes, eerie golden circles, gazed at me implacably, as if nothing else mattered except their instinct to spawn. I reached out, touched one, and then another, and another.

THE EDITOR'S TAKE: This is one of Jodi Peterson's favorite endings because it's personal rather than a summary of the research or a quote from an expert. "[Ray Ring] takes us into the water with him to see and feel what a healthy salmon stream is like," she says. "His ending is intimate, wonderfully descriptive, and brings home in a visceral way the messages of his story."

THE WRITER'S TAKE: Ray Ring, former senior editor at *High Country News*, aims his endings at readers' hearts, not their minds. He also says a great ending is the logical climax to the story:

"It should feel fresh, extending the thrust of the story in some way into surprising territory that still fits with everything that's come before the ending."

Ring starts looking for endings right away: "I'm always scouting for scenes, dialogue, and personal thoughts that might work for the ending." This ending came from a chance opportunity that fulfilled his longtime dream of swimming with salmon. "I figured, if the experience would be powerful for me, it would also be powerful for my readers," he says.

*

If you've followed all this advice but are still stuck on the kicker, remember that even greats like John McPhee can have trouble finding the perfect ending. As he explains in his 2013 *New Yorker* article "Structure," he usually knows the last line from the outset but sometimes has to "struggle for satisfaction at the end."

When that happens, he suggests, "Look back upstream. If you have come to your planned ending and it doesn't seem to be working, run your eye up the page and the page before that. You may see that your best ending is somewhere in there, that you were finished before you thought you were."

McPhee continues: "People often ask how I know when I'm done— not just when I've come to the end, but in all the drafts and revisions and substitutions of one word for another how do I know there is no more to do? When am I done? I just know. I'm lucky that way. What I know is that I can't do any better; someone else might do better, but that's all I can do; so I call it done."

Ending with this advice surely signals that it is McPhee's most important takeaway. And if it's good enough for him, it's good enough for me. So if you take one thing from this post, take this: Treat kickers like the rest of the story—do your best. And when your best is done, so are you.

32 The First Critic Is You: Editing Your Own Work

Tiên Nguyễn

SELF-EDITING IS A selfless endeavor. You cut, replace, rearrange, and endlessly reread—all for the reader's benefit. "Journalism is all about having a sense of empathy with your audience," says Dan Fagin, director of the Science, Health and Environmental Reporting Program at New York University and Pulitzer Prize–winning author of *Toms River: A Story of Science and Salvation*. To achieve that connection with the audience, Fagin says, revise with readers in mind, always asking yourself "what they need, what they want, will they understand?"

As traditional journalism outlets' budgets shrink and editors are being overextended, writers are asked to take increasingly larger roles in shaping their own stories. Whether you have the best editor in the world or no editor at all, self-editing is necessary to deliver the best story possible.

Get Organized

The first step in ensuring clean copy occurs before a word is written: distilling piles of notes and research into a clear structure.

Creating a solid structure need not involve creating a formal outline. Elizabeth Svoboda, freelance journalist and author of *What Makes a Hero? The Surprising Science of Selflessness*, finds a skeletal framework sufficient for establishing a clear structure for a story. "I might jot down just a couple of sentences or phrases that describe each section of the piece," she says. "For example, I might write: 'Intro: Describe scientist wading through wetland to obtain water samples; explain how these samples teach us more about the mix of microbes living in the wetland.' As I actually write each section, I might go back to the outline a couple of times to remind myself of the main points that cannot be lost, but I'm adding a lot more detail on top of those points."

Science journalist Apoorva Mandavilli begins sorting her notes by copying snippets of information and interviews of multiple sources into separate sections of a new document, organized by theme. When she was reporting her recent *Popular Science* feature "The AIDS Cure," for example, most of her sources touched on a similar set of themes. Pasting portions of each interview into different thematic sections helped her find key points related to each theme (and later made it easier to annotate her story for fact-checking).

Virginia Hughes, freelance science journalist and author of the *Only Human* blog at *National Geographic*, takes a similar approach, creating separate documents in Scrivener that each contain a line summarizing one section of the story. As she writes, she says, these sections get "meatier and meatier" until the documents collectively become a draft of the entire piece.

Fagin views transitions from one section to the next as essential to good organization. Transitions should be seamless to keep the reader from slowing down or, worse, stopping. "I want my work to be analogous to one of those moving walkways at the airport, where you just step on and you're moving steadily forward and it's a pleasant experience," he says.

Unleash Your Inner Critic—When You're Ready

Once you begin a draft, it can be hard to tell the processes of writing and editing apart. *Washington Post* science reporter Brian Vastag encourages writers to write the whole story before going back to edit it. Writing a first draft is not like a live musical performance where the audience hears every wrong note, he says. Like a recording artist who can finesse a song until it's flawless, a writer can revise and improve the words on a second go-round.

Fagin says he usually takes a different approach, self-editing as he writes. He cautions his students against being paralyzed by imperfect text, and urges them to continually rewrite sentences until they work. "If you think there's even a chance that it might sound better a different way, try it," he says. "There's nothing easier than abandoning that and going back to what you had before."

As your story progresses, stow away any portions that you cut, perhaps in a separate document. They may or may not make it back into the story, as editors have different opinions about what to do with such material. Mandavilli says that when writers send extras to her, she hardly ever uses them. "If the writers are cutting that stuff themselves, 95 percent of the time it's because it doesn't fit," she says. But extras can come in handy during later conversations with an editor, possibly when talking through alternate approaches to a topic.

Gain Some Distance

Once you've pounded out a workable first draft, find time to step away from your copy for a day, a week, or even longer, so you can come back to it with fresh eyes. Letting your work marinate highlights any problems and makes them easier to cut. "[Taking a break] is valuable every time I do it," says Mark Schrope, a freelance writer and editor who specializes in oceanography.

Distance can also help you realize when you've fallen in love with your own sentences. Tim Folger, a contributing editor at *Discover* and series editor of the annual *Best American Science and Nature Writing* anthology, suggests editing the story as if someone else wrote it. Though it may be difficult, imagine reading it with no prior knowledge and see if any questions arise, agrees Mary Bates, who writes the *Zoologic* blog at *Wired*.

Read Aloud

One of the fastest and most effective ways to tighten copy is to read your piece out loud. Reading your story to yourself reveals missing words, run-on sentences, and unwieldy phrases. Hearing your story also provides a sense of the piece's overall flow and voice.

To emulate the rhythm of a particular publication, read as many back issues as possible. Adhering to its style is important because any missteps draw an editor's attention away from content.

Tune In to Details

Now it's time for line editing. Root out common problems such as passive voice, homophones, misspellings, and repeated words. Try replacing multisyllabic words with shorter, stronger words. Mandavilli recommends removing each word in the sentence one at a time to see if its absence has any impact on the sentence. That rule echoes advice in William Zinsser's classic guidebook, *On Writing Well*: Every word must do work.

The same is true for quotes. When used correctly, quotes engage the reader and make the story come alive. Use direct quotes to convey something remarkable, emotional, contentious, or wise. Most of the time, Fagin says, paraphrasing is adequate for giving information.

Kill Jargon

Remove any words that may confuse the reader. Mandavilli says many science journalists overuse jargon because they are worried others might question their intellect or authority on the subject. "But really, no one has ever complained that something is too easy to understand," she says.

Acronyms are similarly discouraging to readers, says Fagin. Acronyms are tempting because they're convenient for the writer, he says, but remember: It's not about you. Unless a term is used repeatedly throughout the piece, he advises using the original term or a shorter version of it, rather than an acronym. Instead of asking the readers to remember what the DHSEA is, for example, you could describe it as the Dietary Supplement Act.

Though it seems contradictory, self-editing does not have to be solitary. Talk about your story or send a draft to a willing friend or colleague. When the piece involves a controversial issue, Svoboda says she occasionally asks her husband to read it to ensure her point comes across clearly. You might disagree with your outside reader about how to fix a problem, but don't disregard the issue, Fagin advises. Anything that is confusing to the reader should not persist unchanged in the piece, he says.

Adopting these self-editing practices will strengthen your final piece. From major reorganizations to deleting one unnecessary word, every useful edit drives the story forward, Fagin says. And nothing is more important than keeping the reader's interest to the end of your story. "You never ever, in journalism, want the reader to stop."

33 A Conversation with Linda Nordling on "How Decolonization Could Reshape South African Science"

Jeanne Erdmann

WHILE REPORTING FOR *Nature* about the decolonization of science in post-Apartheid South Africa, science journalist Linda Nordling found herself in an uncomfortable spot. Nordling, a native of Sweden, had long lived in the United Kingdom before making South Africa her home more than a decade ago. She was passionately interested in efforts to understand and extinguish the remaining colonial influences and prejudices from South Africa's academic institutions. She believed the story of that ongoing transformation deserved to be told.

But Nordling worried whether it was right for a white journalist from a former colonial power to tell the story—in a magazine that was part of the colonial history of science—of the decolonization of South African academe. She worried that South Africans, including her sources, might accuse her of having appropriated their culture. As Nordling worked through her doubts, she found herself coming up against accepted ideas of objectivity and influence and working harder than she ever had to understand the context of race and culture and power relations.

Nordling's story, "How Decolonization Could Reshape South African Science," was published in *Nature* on February 7, 2018. Here, she shares how she confronted her doubts about becoming the "custodian" of someone else's story.

JEANNE ERDMANN: What made you decide to write this story?

LINDA NORDLING: After I came down here in 2006, I realized you can't live in South Africa without quickly becoming aware about justice, about who gets to speak, and about who gets to tell the story. In the arts and humanities and some social sciences, it's fairly straightforward, but I was struggling with what decolonization means in the sciences. In January 2017, I took a university summer course on decolonization. By then, we'd had about a year of protests in South Africa. At the start, the protests were to remove a statue of the [British] colonizer Cecil John Rhodes from a prominent plinth on the University of Cape Town campus. Once that was successfully achieved, the movement grew to other campuses around the country to include other challenges facing students. Foremost among them were high university fees, but also decolonization. The students called for free, decolonized education.

A few months after the summer course, I attended a seminar on decolonization and Africanizing medicine, and that's when I first heard Wanga Zembe-Mkabile, the social-policy researcher in my lede. She told the story of a project that involved taking food inventories of participants' cupboards, many of whom were embarrassed because there was little or no food in their pantries. I started to feel like there was something there. At this point I decided to pitch a news-analysis story on decolonizing medicine to a publication other than *Nature*. The pitch was accepted.

Wanga Zembe-Mkabile's story seems to have been critical to your reporting. How did you get her on board?

I interviewed Wanga for that story a week or so after the pitch was accepted. At the time, she was concerned about how she would be portrayed, and I said I'd be happy to share a rough draft because it was a condition for her to participate.

So you felt like the ethical imperative to be sure that you were not unfairly co-opting her story was more important than the typical proscription against sharing copy with sources?

That's exactly it. It was a *sine qua non* that she asked for. I did explain to the editor of the news analysis that I wanted to share Wanga's copy because of the sensitivity and personal nature of her story, and the editor said it was fine.

But when I shared the paragraphs I'd written for my draft, [Wanga] did not like it and asked for it to change or to withdraw from the article. Her objections were mainly over tone and turns of phrase like using research "subjects" instead of "participants" or "volunteers," or the word "pitiful" when describing research participants' empty cupboards. To her, this signaled that I was out of touch with the things I was writing about—dignity and decolonization.

That was quite stressful because she was my star witness. It was her story that inspired both the news analysis and eventually the *Nature* feature, and I thought she was at the heart of this story.

I had to go back to the drawing board and think about what she had actually told me. And I realized that I had filled in gaps according to my understanding, and that I had used words that to me made the copy more dramatic—but to her ears sounded disrespectful. It set back the trust I had gained in our face-to-face interview, obviously. So in order to keep her in my story, I had to regain her trust. I entered into a dialogue with her about my use of language. I learnt a lot, not least to listen and to second-guess my instincts.

In the end that news-analysis story was killed, for reasons that had nothing to do with my interactions with Wanga. It was simply that the story I felt I could write was not the story my editor wanted. The pitch was timed on a promise by the South African Medical Research Council to publish a decolonization strategy before the end of the year. That strategy hasn't seemed to have materialized yet, so it was a bad peg at best. The fact that the strategy didn't seem to be as solid a peg as I'd hoped was probably the main reason for the news analysis not working out.

Once that commission was no more, I was free to explore Wanga's story without a particular format in mind. I remember I sat down one afternoon and wrote it like the beginning of a book, or a long magazine profile. I did it because I felt there was so much in it, because I needed to process it—to understand what it was about it that gripped me so much—but also to atone for my first cack-handed attempt at translating it onto the page. I ended up with about 2,000 words of fairly florid prose, and I sent it straight to Wanga. She loved it. That restored my confidence in my ability to write about decolonization.

How did you end up pitching *Nature*?

I never sent Brendan [Maher, a features editor at *Nature*] a proper pitch for the feature. Sometime after the news analysis got canned I talked to him about it on the phone, and he asked me to send what I had. I think

I explained to him all the challenges I had encountered trying to follow the story, and about my fears about whether or not I was the person to write it. I think that all intrigued him. But I got the feeling that he also fell for Wanga's story.

He said to send the notes I had. Besides Wanga's story, I also had snippets from interviews with others that I had conducted for the news analysis. But in my mind, the story had by now evolved into something that wasn't just about medicine, but about research in general. I think it was Wanga's story—especially the quotes I had from her in my long write-up—that enticed him. Things like [Zembe-Mkabile's statement that] "We were fast asleep. At least now, students are alert," [a comment] about the difference between her experience as an undergraduate student at a South African university in the 90s compared with students now.

In that sense, the story as it appeared a few months later in *Nature* had its roots in my interview with Wanga, and I'm not sure it would have had the impact it had if there hadn't been that slip in trust at the very start of the process, and the regaining of it over time.

I did not share the copy for *Nature* [with Wanga], except for checking quotes. Hardly any of that long-form write-up [that I sent her before I sent it to *Nature*] survives in the final *Nature* story. It was just the space in which I proved to Wanga that I did understand her story, and that she could trust me with it.

For the *Nature* piece, how did you get from a story of one researcher's unease with rooting around in people's kitchen cupboards to a more textured story about a movement in South African science? What questions and themes did you want to explore?

When you mention decolonizing science, people usually start with: "How? How do you do it?" That's certainly what drove me to take the summer course in decolonizing knowledge at UCT [University of Cape Town] in early 2017.

I spent every night for a week listening to epistemological explanations and historical accounts about the creation of knowledge, and by the end of it I still was not much closer to an answer. It was obvious how to do it, and why it was needed, in subjects like history or law. But when I asked my lecturer "But how about science?" he got a gleam in his eye. A gleam that suggested to me that many people—perhaps white people like me in particular—had asked the same question of him before, often with the intention to undermine his logic. I bristled at that, because I didn't feel like I had an agenda—I wanted to understand.

Then, when I listened to Wanga speak a few months later, a first puzzle piece fell into place. I felt her story was a key, but not one that unlocked the "how" question. Rather, it was about how it feels to be colonized, and how to engage with part of yourself that was trained under a colonized system.

It took me a while to understand this, but once I did, I decided that instead of chasing examples of how to decolonize science in different disciplines, I could speak to the generation of scientists trained in South Africa who were children under Apartheid, but who trained in science in the early years of democracy. They were not the protesting students of today—but they will carry the brunt of the leadership roles in universities as South Africa transitions into a reality where the majority of researchers are Black, not white.

That gave me a cohort to work with. Their views on decolonization were fresh, and intuitive, not bogged down in epistemological jargon. It felt fresh, and it felt globally relevant, even though it focused on happenings far away from the white-hot centers of global research in Europe, North America, and Asia.

How did you get past your own discomfort about being the right person for this story?

I had huge misgivings about whether I should even be allowed to write this article, not just as a white person but as a white non–South African. For a long time, I left it thinking I was actually not the person, and then I would get dragged back in, and then I'd think, "I can't do this, I'm going to get nailed locally." I now consider this my home. I didn't want to be colonial, and that was a big fear that permeated the entire process and influenced some of the things that I did, such as mind-dumping my interview and sending it to the source, which is something I wouldn't normally do, but I needed that reassurance.

I've spoken about it to many other journalists, and I feel quite a lot of people have the same feelings of anxiety about telling someone else's story—what gives me the right? With this story in particular, it brought it to a head because not only was I writing about Africans, I was writing about race. I was writing about the experience of people and their reaction to a culture that I in some way consider my own, having come from Sweden.

For me that was one of the really big personal challenges, just to feel like I was allowed to give myself permission to write this.

The theme of inclusion and having a voice runs strong in your story. How did you decide what other voices, in addition to Zembe-Mkabile's, to include?

Wanga is a social scientist, and I was keen to get some lab people, but deciding who to speak to was incredibly difficult. For example, I found that few scientists had thought about decolonization, or wanted to engage with it. Perhaps they thought that it was something that didn't apply in the sciences, or they were afraid of saying the wrong thing. Many struggled with the *how* question, like I did. For my purposes I needed people who were happy to open up about how they viewed their careers, their training, their past, and their future. It was quite a big ask, and it required a high level of reflectivity among the people I chose to feature.

I spent a lot of energy and time second-guessing everything, whether I was making a complicated mess and whether I was going to be criticized for omitting certain people and including certain people.

Early on, I decided I wanted the main characters to be Black. I had a friend, one of my sanity readers, go over the story, and she asked why there weren't any white people in it. She thought that was problematic. I think my friend saw in my omitting white people from the article a sort of value judgment, that I was suggesting that the work involved in decolonizing the academy would and should fall to Black people. I was more interested in the experiences of people who went from the experience of not having access to having access. You can totally be into decolonization if you aren't Black—in fact, it's a racially diverse field of study—but I was the most curious about how people belonging to formerly disadvantaged groups under Apartheid, and particularly Black Africans, related to the discussion. I also wanted to make sure I spoke to mostly Black South Africans, rather than scientists from other parts of the continent who may not have grown up in Apartheid.

In the end, I was surprised that white young scientists in South Africa felt the article spoke to them too. It seemed that Wanga's story about belonging, about being an "insider/outsider" scientist [that is, a Black South African who was trained in England and was, as the article notes, "the product of a system shaped by and for white Europeans"], resonated with many young scientists regardless of their provenance. Which might say something about generational shifts and characteristics of young scientists worldwide.

One other thing I wanted to raise in the story is that Black women are the most underrepresented social group in South African science, and the biggest group population-wise—but they hold the smallest share of science decisions. Yet I really ended up with all these women.

Did you intend from the start to include mostly women?

I didn't set out to only find women, and I really struggled with finding South African men who were ready to open up about their personal engagement with decolonization, about belonging in science, about coming from Apartheid, and about navigating the academic space as people of color in new South Africa. (Now I know where I should have looked— there's a Black caucus at UCT that I should have contacted.) I did reach out to a couple of men who have been very involved in decolonization, and they didn't want to talk about it yet. I met [Siyanda] Makaula through a friend at a conference. I was mentioning my story about decolonization, and she went, "Oh, my God, you have to speak with this guy." It was pure serendipity.

What was hardest about putting the story together?

The biggest challenge all the way through was how to structure the story. Brendan and I talked about doing a kind of oral history, which I wasn't quite sure about because I wasn't sure what it meant. Even though we decided to stick to a normal, structured feature with a beginning, middle, and end, thinking about the story as an oral history gave us a bit more space for the personal narratives as something that needed to drive the story. This was very important because the more I tried to pinpoint [what decolonizing science means], the more I felt I was getting sidetracked into emotions. And then it dawned on me that decolonization was more a feeling, rather than something that happened to a place, or to the rewriting of a textbook.

When I started to think about it that way, suddenly I felt like I could write about people's stories and feelings, and that could be enough, rather than to try and pinpoint what decolonization means for geology, or for medicine, or for rewriting curricula, because it's not there yet. Whenever I tried to do that it felt like I was veering off course. I kept having to remind myself that the "how" was intimately connected to the emotions expressed by my sources—that decolonization was something that happens inside. A changed curriculum can help, or having a broader racial representation among teachers and research mentors. But the change, the click, is internal. It's about how you see yourself in the world.

That decision made the structure fall into place.

How did the editing process go?

What was interesting in the editorial process with *Nature* [was that] time and again, people wanted to push it back toward a piece that explained *how* to decolonize science. I had to keep pulling it back. It was easy in that I *couldn't* tell the "how" story! Often I simply had to say, "I can't do more than this." That meant leaving some of the editors hanging.

Still, I actually feel there are some pretty good "hows" in there—for example, when one source says decolonization will happen in the mind, and how it's about what is taught, and by whom, and the examples they give. I think that is pretty concrete. Obviously you could be more granular if you chose a particular discipline, but I don't think there is *no* guidance in the article about how to begin decolonizing science.

What I found really interesting is that there was a generational division in how people viewed the story. Many senior people felt it was too thin on detail on the "how," and that the case studies were chosen because they all work in science areas that are easy to decolonize. But younger people did not seem to see an issue with that at all. They seemed to feel liberated by how it described the interface between science and identity, and the role of the scientist as a social construct, and how that doesn't have to be at odds with doing rigorous science, or require that you have to "start over" with science.

There were also editors who called for the contrary view, from people saying we don't need to decolonize science, which Brendan and I chose not to include. For me it was quite obvious why you might argue that science is universal and doesn't need decolonization, so we didn't include a person just to get the opposing view.

Earlier you mentioned asking a "sanity reader" to go over the story. What was your goal there? And was that a formal sensitivity read that *Nature* encouraged or required?

The sanity check was to make sure none of the nearly finished copy was offensive to a Black South African reader. It was a friend who is a civil servant in the country's science department. It was something *Nature* encouraged, yes. Not required! *Nature* circulated it widely among its editors too for sensitivity input.

Was there anything in the reporting or writing process that you wish you'd done differently?

There were parts of it, especially in the Siyanda [Makaula] interview, that [were] a little legally problematic. Siyanda was very angry, and I wrote the anger, and the editors felt like some of this was a little bit legally difficult. He was talking about specific incidents in his past, which implicated people who could have been identified easily. Part of me regrets that we couldn't reflect more of this anger in the article, but it didn't necessarily drive the story forward. It was more about past injustices, and I can see why the *Nature* editors deemed it not essential. I also think if that part had been essential to the story, it wouldn't have been changed.

How did your own attitudes evolve as you worked on this story?

This article was empowering for me, but it was also very humbling. It made me even more aware of the pitfalls we risk falling into, and what we owe to people whose cultures we don't share when we take on guardianship of their stories and become custodians of their stories. This type of story doesn't translate easily and requires a lot of reflection and engagement.

How has the story been received?

It's been overwhelmingly positive, and not just from Africans but from people of color globally, and also even young white South Africans. Scientists, especially women, have come to me and said that they also feel alienated. The older generation is still very unfulfilled by this story; they still don't quite understand what it means to decolonize science—[for example,] how you do it with molecular biology.

For me, it ended up being positive, even though it was emotionally draining and very difficult. I am still not sure if I am the one to write this. But I went from thinking "why me?" to "why not me?"

PART 5

How Do You Build Expertise in Science Writing?

WHEN I STARTED out in science writing, my questions were basic ones: How do you read a study as a journalist? How do you interview a scientist? How do you convince editors to take a chance on a pitch from a new reporter? As I acquired the basic skills to write a solid science story, my confidence grew—and so did my ambitions. I wanted to tackle more complicated stories, and when I did, I discovered, of course, that there was even more to learn. The first time I reported a story that involved politically controversial issues, I had to learn to bring a more critical mindset to everything I heard. The first time I reported on neutrinos, I had to figure out how to write about subatomic particles in a way that wouldn't melt readers' brains, or my own. And the first time I wrote a story that involved a significant narrative arc, I had to learn how to weave scenes and character development together with the scientific background information that gave the story its reason for being.

None of these skills came automatically or easily. I had help from skilled editors. And I depend on that kind of help to this day. Because for me and for all of us, the learning is never done. You may, for example, find yourself drawn to increasingly complex subjects, some without immediate relevancy for readers. How, for example, do you hook readers who don't know what a quasar is on the discovery that light from one got delayed by seven years as the gravity of galaxies bent its path? How do you explain the challenges of reconstituting RNA from the remains of a now-extinct Tasmanian tiger that died more than a hundred years ago?

On the other extreme, you may find yourself having to convey urgent scientific or medical evidence. For example, how do you investigate and report on maternal mortality rates among Black women, or on the risks your readers face from breathing in wildfire smoke? Or how do you handle the kind of situation so many journalists faced during the COVID-19 pandemic, covering fast-moving science of immediate public importance, when you can't wait out the long, slow process of peer review and publication in scientific journals but instead need to report on preliminary studies that other scientists haven't yet fully vetted? In all these cases, how do you make sense of the numbers, evaluate whether a new set of findings is solid, and—if it deserves coverage at all—convey its significance in a way your readers can understand, feel, and make use of?

Sometimes, scientific studies are fairly straightforward and their implications clear. Other times, you may feel uneasy about a study you are trying to help readers understand, or suspicious that its data may not support its conclusions. How can you dig deeper and spot statistical shadiness? Similarly, you may find a dataset that you're sure is packed with information your readers need to understand. How do you analyze it, make sure your conclusions are right, and convey its significance?

Or you may be reading a piece of science journalism and be wowed by how the writer has tackled one of the issues just mentioned, feeling they've accomplished something you just don't know how to do yet. How do you read it closely, to learn as much as you can from it?

This section gives a taste of some of the many skills you can continue to build as your understanding unfolds. The learning process is eternal. And a core belief that underlies everything we do at *The Open Notebook* is that learning is best done in a supportive community in which we share our skills with one another. I hope that's been evident in the examples throughout this book. Come join us, if you haven't already, to share what you know and learn what you don't yet.

Siri Carpenter

34 **How to Read a Scientific Paper**

Alexandra Witze

IT'S ONE OF the first, and likely most intimidating, assignments for a fledgling science reporter. "Here," your editor says. "Write up this paper that's coming out in *Science* this week." And suddenly you're staring at an impenetrable PDF—pages of scientific jargon that you're supposed to understand, interview the author and outside commenters about, and describe in ordinary English to ordinary readers.

Fear not! *The Open Notebook* is here with a primer on how to read a scientific paper. These tips and tricks will work whether you're covering developmental biology or deep-space exploration. The key is to familiarize yourself with the framework in which scientists describe their discoveries, and to not let yourself get bogged down in detail as you're trying to understand the overarching point of it all. (See an illustrative markup of a *Science* paper at bit.ly/TONpapermarkup.)

But first, let's break down what a typical scientific paper contains. Most include these basic sections, usually in this order:

The **author list** is as it sounds, a roster of the scientists involved in the discovery. But hidden within the names are clues that will help you navigate the politics of reporting the story. The first name in the list is often (but not always) the person who did the most work, perhaps the graduate student or postdoc who is the lead on the project. This person is usually (but not always) designated as the "corresponding author" by an asterisk by their name, or by their email address being given on the first or last page of the paper. If the corresponding author is not the first name in the author list, then take extra care to Google the various authors and figure out how they relate to one another. (In many fields, such as biology and psychology, the last author in the list is typically the senior author or lab head. In others, such as experimental physics, where the author list can number in the dozens or hundreds, authors are usually listed alphabetically.) The senior author might be able to provide some broad perspective as to why and how the study was undertaken. But the first or corresponding author is much more likely to be the person who actually did the work, and therefore your better request for an interview.

The **abstract** is a summary of the paper's conclusions. Always read this first, several times over. Usually the significance of the paper will be laid out here, albeit in technical terms. A good abstract will summarize what research was undertaken, what the scientists found, and why it's important. Relevant numbers such as the statistical significance of the finding are often highlighted here as well. Abstracts are prone to typographical errors, so be sure to double-check numbers against the body of the paper as well as your interview with the author.

The **body** of the paper lays out the bulk of the scientific findings. Pay special attention to the first couple of paragraphs, which often serve as an introduction, describing previous research in the field and why the new work is important. This is an excellent place to hunt for references to other papers that can serve as your guidepost for outside commenters (more on that later). Next will come the details of how the research was done; sometimes much of this is broken out into a later **methods** section (see below). Then come the **results**, which may be lengthy. Look for phrases such as "we concluded" to clue you in to their most important points. If statistics are involved, see Rachel Zamzow's primer on how to spot shady statistics.

The final section (sometimes labeled as **discussion**) often summarizes the new findings, puts them in context, and describes the likely next steps to be taken. If your reading has been dragging through the results section, now is the time to refocus. "That sort of information will help a writer answer the nearly inevitable 'so what?' question for their readers as well as their editors," says Sid Perkins, a freelance science writer in Crossville, Tennessee, who writes for outlets including *Science* and *Science News for Students*.

The **figures** are the data [tables], graphics, or other visual representations of the discovery. Read these and their captions carefully, as they often contain the bulk of the new findings. If you don't understand the figures, ask the scientist to walk you through them during your interview. Don't be afraid to say things like, "I don't understand what the x-axis means."

The **references** are your portal into a world of additional inscrutable PDFs. You need to plow through at least a couple of the citations, because they are your initial guide to figuring out who you need to call for outside comment. The references are referenced (usually by number) within the body of the text, so you can pinpoint the ones that will be most helpful. For instance, if the text talks about how previous studies have found the opposite of this new one, go look up the cited references, because those authors would be excellent outside commenters. If you do not have access to the journals described in the references, you can at least look at the papers' abstracts, which are always outside the paywall, to get a sense of what those earlier studies concluded. (For further caveats on references, see below.)

The **acknowledgments** are meant for transparency, to show the contributions of the various authors and where they got their funding from. Things to look for here are whether they thank other scientists for "discussions" or "review" of the work; sometimes peer reviewers are explicitly acknowledged as such, in which case you can call those people right away for outside comment. Occasionally there are humorous tidbits that you can pick up on for a story, such as when authors thank the field-camp guards who kept them safe from predatory polar bears. The funding section is usually pro forma, but it is worth scanning for mention of unusual sources of income, such as from a science-loving philanthropist. If the authors declare competing financial interests (such as a patent filing) you will need to report those out and make sure you understand what financial conflicts of interest may be clouding their objectivity.

The **methods** often appear in a ridiculously small typeface after the body of the paper. These lay out how the actual experiments were done. Scour these for any details that will bring your story to life. For instance, they might describe how the climate models were so complicated that they took more than a year to run on one of the world's most powerful supercomputers.

Supplementary information comes with some but not all papers. In most cases it is extra material that the journal did not want to devote space to describing in the paper itself. Always check it out, because there may be hidden gems. In a 2015 study of global lake warming, the only way to find out which specific lakes were warming—and talk about the nearest ones for readers—was to wade through the supplementary information. In another recent example, Harvard researchers left it to the supplementary information to explain that they cranked up a leaf-blower to see how lizards fared during hurricanes, a fact that the Associated Press's Seth Borenstein turned into his lede.

So now you're armed with the basics of what makes up a science paper. How should you tackle reading for your next assignment? The task will be more manageable if you break it into a series of jobs.

Strategize during the First Pass

Your first dive into a paper should be aimed at gathering the most important information for your story—that is, what the research found and why anyone should care. For that, consider following the approach of Mark Peplow, a freelance science journalist in Cambridge, England, who writes for publications including *Nature* and *Chemical & Engineering News*.

If it's a field he's relatively familiar with, such as chemistry or materials science, Peplow takes a first pass through the paper, underlining with a red pen all the facts that are likely to make it into his initial draft. "That means I can produce a skeleton first draft of the story by simply writing a series of sentences containing what I've underlined, and then go into editing mode to jigsaw them into the right order," he says.

As Peplow reads, he looks for numbers to help make the story sing (". . . so porous that a chunk of material the size of a sugar cube contains the surface area of 17 tennis courts"; in the annotated paper, I've flagged intriguing numbers in orange highlighter) and methodological details that might prompt a fun interview question ("How scary was it to be pouring that very hazardous liquid into another one?"). He also keeps an eye out for anything indicating an emerging trend or other examples of the same phenomenon, which can be useful for context within the story or as a forward-looking kicker.

But what if the paper is in a field you're not experienced with, and you don't understand the terminology? Peplow has a plan for that too. "I read the abstract, bathe in my lack of understanding, and mentally throw the abstract away," he says.

Then he goes through the paper, underlining fragments he understands and putting wiggly lines next to paragraphs that he thinks sound important, but doesn't actually know what they mean. Jargon words get circled, and equations ignored. He forges onward, paying attention to phrases such as "our findings," "revealed," "established," or "our measurements show"—signs that these are the new and important bits. "Once I've reached the end of the paper, and I'm sure I don't understand it, I remind myself it's not my fault," Peplow says.

At that point, Peplow starts looking up definitions for the jargon words, either with Google or Wikipedia or in a stack of science reference books he picked up for free when a local library closed. He jots definitions of the words on the paper. To understand concepts, he sometimes searches EurekAlert! for past press releases that explain core concepts, or Googles a string of keywords and adds "review" to hunt for a more comprehensible description. By this point, Peplow can circle back to the paragraphs marked with wiggly lines and start to understand them better. What he doesn't yet comprehend, he marks down as an interview question for the researcher.

Circle Back for What You May Have Missed

Before picking up the phone for that interview, it's worth making a second pass through the paper to see what else you need to help you in your reporting. Check, usually near the end of the paper, to see whether the scientists discuss what the next steps should be—either for their own team or for other groups following up to confirm or expand on the new results, says Perkins. That can provide a ready-made kicker for your story.

Susan Milius, a reporter who covers the life sciences for *Science News*, often makes a beeline straight for the references to try to start identifying outside commenters for a piece. She will find those PDFs and then look within the references' references to build a broad understanding of the field. One caveat, though: Be sure to research how these possible commenters are connected to the author of the current study. Once, Milius phoned an outside commenter who had published on the topic in question some years earlier—but that scientist turned out to be the spouse of the new paper's author. She had a different last name than her husband.

It's also worth remembering that the authors may well be biased in which references they include in the paper. Self-citations, in which authors try to boost their citation count by adding their previous publications to the reference list, are common. And sometimes authors deliberately omit papers by competing groups, a fact that is not always caught during the peer-review process. So don't rely on the references within the PDF to be comprehensive; try a Google Scholar search using keywords from the paper to unearth whether there are competing groups out there.

Other clues may lie in how long the manuscript took to make it through the peer-review process. For many journals these dates come at the very end of the paper, marked something like "submitted" and "accepted." Different journals have different timescales for publishing, but it is always worth looking to see whether the manuscript languished an extraordinary amount of time (like many months) in the review process. If so, ask the author why things took so long. (A fairly innocuous way to do this is to say something like, "I noticed it took a while for this paper to be accepted. Can you tell me how that process went?" Then be prepared for the authors to go on a rant about peer review.)

Hunt for Extra Details

Finally, see if there are additional sources of information you can sweep into your reporting. Check to see if the author's institution is issuing a press release about the work; if this isn't already posted on EurekAlert!, ask the author during the interview if they are preparing additional press materials and, if so, how you can get hold of those. This is also a good time to ask for any art, such as photos or videos to illustrate your story. You will of course have already looked at all their figures in detail, so you'll be well placed to request the art that is most relevant to what you and your editor are looking for.

With these tools at your side, you should be well suited to tackle your next scientific paper.

35 What Are the Odds? Reporting on Risk

Jane C. Hu

IF A PREGNANT woman visits Miami next summer, how much risk of contracting the Zika virus will she face? What is the risk that rising oceans will put New York City's streets under water by 2100? Is coffee more likely to benefit your health or to harm you? What factors predict whether a child will graduate from high school?

Risk is at the heart of many of the health, environmental, and science stories people care about most. People want to know how likely they are to be helped by a drug, contract a virus, get cancer, find themselves in the path of a hurricane, or experience a catastrophic earthquake—and they want to know how they can influence their odds. Sometimes risks and the ways of minimizing them come with concrete numbers—perhaps taking medication X decreases a person's chances of dying from disease Y by Z percent. Other times, writing about risk is less straightforward. Stories often involve more abstract representations of risk, like whether a particular intervention improves happiness or whether using a smartphone affects a person's attention span.

Clear, accurate, and thorough reporting should always be a journalist's foremost goal. Those qualities are all the more crucial when writing about risk, because readers use the products of risk reporting to inform their own decision-making.

Here are some tips to help you report on risk with precision and clarity.

Describe Risk in the Most Meaningful Way

Not all expressions of risk are equally easy to understand, and two equally accurate descriptions of a given risk can have very different psychological effects on readers.

Consider, for example, a 2012 study that found that women who suffer migraines were 40 percent more likely to develop multiple sclerosis (MS) than women who did not have migraines. That's a description of what is known as *relative risk*: one group's risk compared to another's. But presenting only relative risk omits important context and can thereby make the stakes seem more extreme. A 40 percent increase in risk sounds like a lot. But consider the two groups' *absolute risk*: their likelihood of developing MS in the first place. According to the 2012 study, about 32 in 10,000 women who don't have migraines develop MS; for women who do have migraines, the absolute risk is about 47 in 10,000. When expressed in this way, the two groups' difference in risk doesn't seem as large.

Dispel Misconceptions

Science journalists can help orient readers by identifying popular misconceptions about issues and explaining what scientists actually know. For example, many consumers are understandably concerned about the safety of chemical food additives. Some food additives do pose known or suspected health threats. But public misunderstandings about chemicals in foods are common.

In some cases, bloggers and other media figures have contributed to such misunderstandings. Vani Hari, who writes the blog *Food Babe*, is infamous for launching public campaigns against food additives. Hari argues that any chemical found in a nonfood product should not be used in food. In 2014, she successfully petitioned the fast-food chain Subway to remove azodicarbonamide, a flour-whitening agent, from its bread, arguing that the chemical is unsafe because it is also found in yoga mats. But according to the FDA, azodicarbonamide is safe. Reporters should remind readers that all foods contain chemicals, and that an unfamiliar chemical name doesn't necessarily signal a health hazard.

Turn on Your Bullshit Detector

In researching and reporting stories—even short news stories—make sure your bullshit detector is on. Be critical of where your information comes from and be alert to potential biases among your sources. When evaluating findings disseminated by corporations, in particular, it's essential to evaluate claims carefully. Companies often sponsor research and report their findings via press releases rather than through peer-reviewed journal articles. Some even produce white papers that look like snazzier versions of academic papers. These white papers may contain statistics, graphs, or even original findings like a peer-reviewed paper, but beware: The ultimate goal of corporate white papers is to make the company's own products or treatments look good.

To avoid being taken in, always check study authors' affiliations and, when reading press releases, look for mention of what journal the study is published in. "Following the money is always useful," says Olivia Solon, a freelance science and technology journalist who writes for the *Guardian* and *Wired*.

Solon recently wrote for *Mosaic* about the effects of children playing with smartphones and tablets, and while doing her research she came across many papers on the websites of e-learning tools and apps that were funded by the companies themselves. These papers, which were not peer-reviewed, typically reported more tech-friendly results. When Solon turned to sources that were not affiliated with children's-app companies, she found a different story: There's scant research-based evidence that children can learn from these apps.

Academics, too, have a stake in the game, as do academic journal publishers, meeting organizers, funding agencies, and university news offices. Sure, most academic scientists are primarily motivated by the quest for knowledge. But researchers and the organizations that support them also benefit from favorable press coverage. Be wary of researchers who push you to get the word out about their "exciting work," and recognize that not all published research is necessarily *good* research—predatory and otherwise bogus journals abound. (For instance, Elsevier has a journal called *Homeopathy* among its offerings.)

Talk to Independent Experts

Talking with a variety of expert sources is a staple of high-quality science reporting, and for good reason—it's an excellent way to get needed background, correct misconceptions, and refine your understanding of the big picture. This is especially important for stories that require a delicate touch, such as stories about the risk of an epidemic or the risks associated with a controversial medical treatment. "When there's fear about stuff, you've got to be extra careful," says Solon.

She recently came across a study showing how virtual reality games might reduce obesity, but she was wary of writing an article on the basis of just one study. So she sought comment from clinical experts who were not involved in the research, and they told her that the treatment would only be effective for a very small subset of patients. "Had I not spoken to them, I could have written a piece that made out that virtual reality would be a solution for obesity," she says. "It was a good *story*, but there just wasn't enough real evidence." She decided to can the story.

Helen Branswell, an infectious-diseases and public health reporter at *STAT*, says she has a rolodex of epidemiologists, biologists, doctors, and other experts to turn to when she's unsure about a new story. "You can call them up and ask, 'Hey, does this seem real? What do you make of this?'"

In addition to asking general questions, ask experts specific questions about research methods or statistical analyses you're not familiar with, and check your own knowledge by asking experts if your understanding of the topic is accurate. For instance, you can try something like, "So, azodicarbonamide *is* found in both breads and yoga mats, but that's fine? Why is that fine?" Questions like this can help you make sure your reporting is airtight, and they may even prompt your interviewer to reveal nuances or new information that further strengthen your understanding of the issue.

At the same time, Branswell cautions, beware the "science experts" who feel they can speak about anything under the sun. "Just because someone knows something about respiratory diseases doesn't mean they're the right person to talk to about antibiotic resistance," she says.

Remember that you can (and should) get independent experts' opinions on new studies or reports—even on those that are under a news embargo. Just be sure the expert understands the embargo and agrees to honor it. (Any organization or journal that forbids you from getting outside comment on a news story is one to be wary of.)

Be Clear about the Unknowns

Scientists shy away from using verbs like "prove" to describe their work. Reporters should, too. While readers may want definitive takeaways, stories about risk don't always lend themselves to neat narratives. Sometimes the best reporting indicates what we *don't* know—and then follows up with another story as the missing pieces fall into place. "It's really important not to overstate; if the facts are uncertain, then you should convey that," says Branswell. "We do readers a disservice if we portray things with too much certainty."

With new discoveries made everyday, science news can change quickly. [In mid-2016], for example, epidemiologists thought the only mosquito species that carried the devastating Zika virus were *Aedes aegypti* and *Aedes albopictus*, and that men could infect female sexual partners but not the reverse. New research has found that both these theories were incomplete, and that Zika may be even more spreadable than previously thought: *Culex* mosquitos, which are more commonly found than *Aedes*, can also carry Zika, and women *can* infect men. That earlier reports turned out to be incomplete is no fault of scientists or reporters, of course; it's just the way science works.

36 **Spotting Shady Statistics**

Rachel Zamzow

WHEN FREELANCE SCIENCE journalist Tara Haelle first started writing about medical studies, she admits she had no clue what she was doing. She quoted press releases and read only study abstracts—practices that make most science journalists cringe. But in the spring of 2012, Haelle attended a workshop run by health journalism veterans Gary Schwitzer and Ivan Oransky at the annual conference of the Association of Health Care Journalists (AHCJ) in Atlanta. The two watchdog powerhouses—Schwitzer spearheads *Health News Review* and Oransky, *Retraction Watch*—taught Haelle and the other attendees how to catch flaws in research studies. Haelle came away from the session a new journalist. "I was like, 'Holy crap, I've been doing it wrong,'" she says.

For years after the workshop, Haelle put those new skills into practice, poring over research studies for hours and delving deep into the thicket of their methods and results sections. Now the AHCJ Medical Studies Core Topic leader, she regularly takes down overblown or shoddy science. Some of the issues Haelle calls out involve questionable practices like excessive data mining or cherry-picking subjects—activities that likely reflect increasing pressure on scientists to produce eye-catching results.

The problem, according to Haelle and others, is that this statistical fudging has grown harder to detect. Much of it is buried in the way data are sliced and diced after the fact or put through tortured analysis in a search for significant results. The good news is that science journalists can learn to catch many of these tricks without getting a degree in biostatistics, Haelle says. "You don't have to be able to conduct your own regression analysis to be able to look for problems."

As journalists comb through a study's nuts and bolts—avoiding the temptation to skim through the dense prose of the methods and results sections—they should also look out for specific red flags.

Data Dredging

One of the more common ways researchers might try to tweak their results is by a practice known as p-hacking, which entails mining a dataset until you get a finding that passes the bar of statistical significance. This is determined by a p-value—a measure of how unexpected a finding is, if in fact a researcher's hypothesis were wrong. (Most scientific fields consider a p-value of less than .05 statistically significant.)

P-hacking takes advantage of the flexibility researchers have to massage their data after it has already been collected, according to biostatistician Susan Wei at the University of Minnesota. In most research studies, researchers must make judgment calls about which of several analytical approaches to use, says Wei. She notes that you could give the same dataset to five statisticians, and each would come up with a different result. "Data analysis is not quite a science," she says. "There's a bit more art to it."

P-hacking can take a few different forms, Wei says. In some cases, researchers might test for relationships between a bunch of variables—say, between different doses of an antidepressant and their effects on a variety of different outcomes, such as mood, appetite, sleep patterns, or suicidal thoughts. If a significant result pops out for any of those outcomes, the researchers may then generate a hypothesis to fit that finding, ignoring those variables for which there was no effect. Coming up with a hypothesis retroactively—sometimes referred to as "hypothesizing after the results are known," or HARKing—often follows this form of p-hacking.

Another way researchers can p-hack data is to stick with their original hypothesis for a study, but cycle through several statistical tests until they achieve the result they were hoping to find. For example, if a study includes patients' ages as a variable, researchers could first use a statistical technique that treats age as a continuum. If this approach doesn't produce a significant finding, they could try a different analysis that breaks the ages into categories, such as people younger than 50 and those who are 50 and older.

Both of these tricks increase the likelihood that a significant result is actually just due to plain chance. Since researchers typically consider

p-values less than .05 statistically significant, running 20 or more tests almost ensures that at least one finding is a false positive. As a result, researchers may be drawing connections that don't actually exist. In a March 2017 blog post for AHCJ, Haelle pointed out p-hacking in a study that found an association between certain vaccines and psychiatric conditions, including obsessive compulsive disorder and anorexia. As Haelle noted, the researchers ran an enormous number of tests based on different combinations of vaccines, administration time points, and psychiatric conditions, making it nearly inevitable that some significant result would arise by chance alone.

Spotting p-hacking can be difficult, Wei says, because researchers may sanitize their final publications by only reporting the tests that produced significant findings. But reporters can look for evidence that it might have occurred by asking researchers if they examined other variables or used extra statistical tests that weren't listed in the paper—and if so, ask how they settled on the published ones.

The methods and results sections of studies can reveal other signs of p-hacking. For example, Haelle says, watch out for an excessive number of endpoints, otherwise known as dependent variables. Clinical studies should generally have one or two primary endpoints and maybe four to six secondary ones. Much more than that—even if all the endpoints are reported—signals that the researchers might have gone fishing for a statistically significant result. Haelle also scans papers for mentions of a Bonferroni correction, which is one of a few ways to account for a large number of statistical tests or endpoints. If she doesn't find a correction when she thinks one was warranted, she asks the researchers why, or if they accounted for multiple tests with a different technique.

Also, examine the published p-values themselves. If they all tend to land near .05, that's a sign that p-hacking might be in play, says Brian Nosek, a psychologist at the University of Virginia and executive director of the Center for Open Science, which encourages transparency in science. P-hacking practices tend to nudge nonsignificant p-values to just under .05. Highly significant findings, such as those with p-values less than .01, are less likely to have arisen from p-hacking.

To verify that a medical study hasn't been p-hacked, check its registration on ClinicalTrials.gov. In most study records, researchers specify the endpoints they intend to use before a project begins. The tabular view of a record lists both the current and original endpoints, as well as a link to older versions of the same study. Researchers can also opt to preregister studies in any field at the Center for Open Science's Open Science Framework, where they can submit detailed plans for data collection and analysis up front. Some 7,000 studies are now preregistered on the site, Nosek says.

These public records allow journalists to compare the original plan for a study to the published paper. Discrepancies are surprisingly common, according to the results of a project run by Ben Goldacre, senior clinical research fellow at the Centre for Evidence-Based Medicine at the University of Oxford in England. Goldacre and his team checked the reported outcomes of clinical trials published in five top medical journals from October 2015 to January 2016 against their initial protocols or public registries. Looking at 67 trials, they found that on average, each reported only around 58 percent of its original outcomes in the published paper and added about five new outcomes.

Lastly, p-hacking may deliver findings that, while statistically significant, aren't that meaningful or of much practical use. What's more, p-values in general aren't even designed to gauge the strength or importance of a finding. Instead look at effect sizes, says Regina Nuzzo, a freelance journalist and statistics professor at Gallaudet University in Washington, DC. Effect sizes can show *how* different two groups are, for example, instead of just reporting that there is a difference.

To find information about effect sizes within studies, search for keywords like "Cohen's d" or "r-squared," which are standard measures of effect size. And ask researchers to interpret their effect sizes in the context of the real world—for example, how much a drug actually helps people with chronic obstructive pulmonary disease.

If an effect size is very small or missing from the results altogether, that's a warning sign that needs following up, Nuzzo says.

Taking an Early Peek

In a practice called data peeking, or interim analysis, researchers may analyze some of their data before it's all collected. Doing so may help a scientist get an early look at whether a drug is working, for example, but it's a bad idea statistically, says Nosek. Running some initial tests on data and then again at the end of a study carries the same pitfalls as *p*-hacking, he says, because the data undergoes multiple rounds of statistical tests.

Data peeking is hard to detect, unless a study was preregistered, because researchers rarely disclose it, Nosek says. However, checking the results of a study early may motivate a scientist to switch out endpoints or even adjust an intervention midstream. Tweaking an intervention before a study ends is especially tricky, says Wei, because when that occurs, it's impossible to say whether the intervention being tested really worked, or if the changes themselves nudged the results toward significance.

And if a study is halted after interim analyses—for example, a drug trial is stopped early due to safety concerns—remember that the findings are less reliable, says Lisa Schwartz, a professor of medicine at the Dartmouth Institute for Health Policy and Clinical Practice in New Hampshire. Researchers may even end trials early if a drug or therapy seems to be overwhelmingly beneficial. But treatment effects can vary over time, Schwartz says, so what seems like a positive finding at the time of data peeking could level out by the end of a study.

If a published paper lists analyses that were run before all of the data were collected, Schwartz advises that journalists ask researchers if these early checks were preplanned in the study's original protocol. (Sometimes this is part of quality control.) "The answers to that question start to help you gauge your level of worry about whether this is really something that is in the suspect-science category," she says.

Slicing and Dicing

When statistical analyses based on an entire dataset don't pan out, researchers may choose to cut out a part of the data for a separate analysis. Perhaps a treatment only benefits women, or maybe a drug is most effective in children younger than a certain age. Analyzing a subgroup of collected data is a very common practice, especially in the context of clinical trials, where drugs are rarely panaceas, says Wei. But this practice is not without downsides. "Subgroup analysis has always been rather controversial," she says, even though almost everyone does it.

The problem with looking at a subset of data is that the results from these analyses are often not reproducible, Wei says. In other words: If a subgroup of patients seems to be particularly responsive to a drug, researchers should then verify the result by conducting a whole new study using only people with the characteristics of the original subgroup. But chances are that this follow-up study won't come up with the same findings. This may happen if the number of participants in the initial subgroup is small, making the results less reliable. Findings that turn out to be significant are likely just false positives. For example, a 2011 study of the fragile X drug mavoglurant showed benefits for a subgroup of seven people with the autism-related condition who had a specific genetic profile. But subsequent studies of the drug didn't reveal the same results.

However, if the original study contained a large number of participants, subgroup analyses might signal a real result. For instance, when researchers broke down data—by age and time since menopause—from over 27,000 postmenopausal women participating in two large Women's Health Initiative trials, they found that hormone replacement therapy can have dramatically different effects and risks. Younger women in the early stages of menopause may be able to take hormones to relieve symptoms like hot flashes, but older women using the therapy long-term have a higher risk of breast cancer and other illnesses.

When reading research articles, it's usually easy to see when scientists have pulled out a chunk of their data for a closer look, Haelle says. For example, after listing the results for the full dataset, a paper may include findings based on subsets of the data, such as participants of a certain gender or race. In those cases, probe further, and ask whether the subgroup being tested makes logical sense considering what's being studied.

Christie Aschwanden, lead science writer at *FiveThirtyEight*, cautions that reporters should be careful of "just-so stories" (alluding to Rudyard Kipling's collection of fantastical stories about how certain phenomena came to be). Scientists may craft a narrative to justify testing a subgroup after they find that it produced a significant result. "These stories perfectly match the data because they were created for the data," she says.

Also, keep in mind that researchers might test several smaller portions of their data until they see a positive effect, which is a form of *p*-hacking, Wei says. "You're fishing until you catch a fish," she says. "Of course you're always going to catch one if you keep going."

Journalists should be particularly wary of studies that test a slew of subgroups of people with specific genetic characteristics, says Dartmouth's Schwartz. Projects like these often have large datasets, which means researchers can test an immense number of correlations between different genetic traits and diagnoses. These studies and other types of subgroup analyses are important for screening ideas, Schwartz says: They help scientists formulate hypotheses for future studies. But they do not offer a reliable result by themselves.

Picking a Proxy

Researchers also have flexibility in choosing their study's endpoints, some of which are more useful than others. When studying a drug designed to prevent heart attacks, for example, a scientist may measure blood pressure or cholesterol as a proxy for the number of heart attacks that the drug purportedly prevents. These so-called surrogate endpoints save time and money when other variables, like deaths, could take many years and millions of dollars to measure. This is mostly a legitimate practice, Wei says, since there are typically scientific standards for what makes an endpoint a good proxy to use.

Some surrogate endpoints are even accepted by the US Food and Drug Administration as grounds for drug approval, says Schwartz. Still, read these with caution. For instance, researchers have tested diabetes drugs by measuring changes in average blood sugar levels instead of more direct measures of patient well-being. But issues with the drug Avandia called the use of this marker into question. Even though their average blood sugar decreased, people taking the diabetes drug had an increased risk of heart attacks and other cardiovascular complications.

"Surrogates always introduce [a] leap of faith," says Schwartz. You hope they translate into outcomes that make a difference in people's lives, but it's not a guarantee. This is why studies using surrogate outcomes deserve a wary eye.

The use of biomarkers in psychology research is another example of potentially flimsy surrogates, says Haelle. For instance, she notes, researchers may measure certain types of brain activity as an endpoint in a drug study involving people with schizophrenia. But these measures aren't yet reliable as markers of actual treatment response. Keep in mind that researchers could be reporting a surrogate endpoint simply because it revealed a significant finding, when other more meaningful endpoints did not. Measures of quality of life are examples of endpoints that may better mark a patient's overall outcome.

Additionally, researchers may roll several endpoints into one composite outcome, which allows them to test different combinations of endpoints—number of heart attacks, hospital admissions, and deaths, for example—until something turns up significant. But these results are vague when it comes to interpreting how a treatment may actually improve patients' lives. For that reason, journalists should check the validity of surrogate outcomes with outside sources. Haelle notes that scientists who are also practicing physicians are especially helpful, because they understand the importance of measuring outcomes that are meaningful to patients.

Don't Go It Alone

With enough practice, journalists can learn to spot some instances of *p*-hacking or other statistical sketchiness. But for many research articles, wading through the statistical morass is not easy. "This is a big ask for journalists to be able to identify these things," says Nosek. After all, if a questionable act made it into a publication, that means it got by the other researchers and journal editors who reviewed the paper. Thus, it's smart reporting—not a failure—to reach out for help when you need it.

A few organizations provide resources to help journalists brush up on common statistical errors and sketchy tactics. *Health News Review* publishes a toolkit of tips for analyzing studies, such as what it really means (and doesn't) if a finding is statistically significant. The AHCJ has several tip sheets on covering medical studies. (Some are available only to members of AHCJ.) The association also has related sessions, like the one Haelle attended and now helps run, at its annual conference, as well as at regional workshops across the country.

Science journalists should also develop relationships with experts who can help decipher the complex methods and results of papers. "Make friends with a statistician, buy them some beers, and put their number on speed dial," says Nuzzo. Additionally, the resource STATS .org aims to improve statistical literacy through workshops at news organizations and through its STATScheck service, which journalists can use to submit questions to statisticians, including Nuzzo. Reporters can also reach out to the American Statistical Association, which can identify statisticians who are familiar with the particular field a journalist is covering. For example, the association helped Aschwanden connect with an expert who could help her understand the complicated methods of cloud seeding for a story she was working on.

In the end, journalists should be careful not to assume that the appearance of shady statistics in a paper makes the researcher an outright fraud. Generally, scientists are motivated to discover truth and find good evidence to back it up. "Mostly the questionable research practices are done out of ignorance, not out of deliberate intent to deceive," says Nosek.

Part of the reason problems like p-hacking and data peeking have arisen, Wei says, is that traditional statistical techniques are no longer relevant for increasingly complex study designs. "We're aware of the issues," she says. "We're trying to give scientists tools to do more rigorous science, which hopefully leads to better reproducibility."

As Aschwanden made clear in her [2015] award-winning *FiveThirty Eight* feature "Science Isn't Broken," science is messy—and just plain hard. And scientists themselves aren't perfect. Just like anyone else, she notes, they are susceptible to the natural bias that drives us to think we're doing good work even when we aren't. "Of course a researcher is going to have a tendency to want to overplay their findings," she says. "This is a human impulse, and it takes work to overcome this." And that work doesn't stop with scientists—it also extends to the journalists who cover science.

37 Problems with Preprints: Covering Rough-Draft Manuscripts Responsibly

Roxanne Khamsi

AT THE END of January [2020], as the new coronavirus raged in Wuhan, China, and cases cropped up in other countries, a rough-draft version of a scientific manuscript went online in which researchers posited that the new pathogen might have acquired portions of its RNA sequence from HIV, perhaps making it more infectious. A Reuters analysis found that the study was tweeted more than 17,000 times and was picked up by at least 25 news outlets in the days that followed. But the takeaway from this manuscript was not related to its conclusions, which proved unfounded. The real lesson was about how reporters should be wary of such unvetted scientific reports: Just two days after the preliminary paper went online—after other scientists noted fundamental flaws—it was withdrawn.

The episode was an early indication that rough-draft scientific manuscripts, known as preprints, would feature heavily in this pandemic, as would concerns about their rigor. In a more recent example, an unvetted genetic analysis suggesting that a mutant version of the coronavirus had evolved to become more contagious was found to be overblown, and journalism watchdogs questioned its coverage. And in yet another case, a study using antibody tests on volunteers in Santa Clara county in California claimed that 50 to 85 times more people in the area had been infected with SARS-CoV-2 than previously thought, suggesting a lower risk of death from the virus. But the study, which had been posted as a preprint, was quickly panned by many epidemiologists who pointed to significant problems, including how the researchers had recruited individuals for testing. A revised version of the manuscript appeared on medRxiv two weeks later. It has yet to appear in a scientific journal.

Preprints are manuscripts that are made publicly accessible online after experiments are done but before undergoing a coordinated review by experts in the relevant field of study to assure that the conclusions are supported by the data. The first preprint server, called arXiv and devoted to physics papers, was launched in 1991 and remains in use today. The rise of preprints in biology and medicine is more recent; the well-known life-sciences preprint servers bioRxiv and medRxiv were launched in 2013 and 2019, respectively. A paper published in April—still in preprint form itself—found 44 platforms for preprints that are biomedical and medical in scope. "The world is moving to a preprint-first world," says Ivan Oransky, cofounder of *Retraction Watch*.

The rise of preprints in biology and health is staggering. Even though only around 2.3 percent of published papers indexed in the PubMed database appear in preprint form before publication, the number of preprints in the biological sciences has gone up more than threefold in the last two years, from around 2,500 per month to around 8,000 a month.

The COVID-19 pandemic has further boosted the visibility of preliminary, unvetted manuscripts. In a newly posted manuscript—itself a preprint that the researchers have not yet submitted for review—Jonny Coates, a postdoctoral researcher in immunology at Cambridge University, and his colleagues compared traffic metrics for all studies posted on bioRxiv and medRxiv between September 2019 and April 2020. Preprints pertaining to COVID-19, all of which were posted this year, were viewed more than 15 times as often as non–COVID-19 preprints. (Some preprint servers, including bioRxiv, flag manuscripts related to SARS-CoV-2 with a note reminding users that preprints are preliminary and should not be treated as conclusive.)

Media coverage of preprints has also risen markedly since the emergence of SARS-CoV-2. That increase is not surprising, says Lauren Morello, deputy health care editor at *Politico*. "The pandemic is accelerating a trend, but this was coming already," she says.

But press coverage of preprints is fraught. Hastily conducted and reported scientific studies are an unfortunate hallmark of the current pandemic, as journalist Christie Aschwanden wrote recently in *Wired*. She says journalists should have their guard up. "Where I've seen reporters go wrong on this is when they sort of grab these [preprints] because they want to be first," Aschwanden says. "We do have pressure to be first and to break news, and I think preprints seem to offer this shiny opportunity to do that. But most of the time that's not really going to pan out." Reporters, she says, might even benefit by assuming any given preprint is "probably a false lead."

A Skeptical Eye

Perhaps the most central question for journalists is whether or not to even cover a preprint in the first place. In some cases, it might be worth covering a low-quality preprint that's going viral on social media in order to shine a sobering light on its flaws. However, journalists need to be cautious about putting preprints in the spotlight. "Even if you are putting all of these qualifiers in and [saying,] 'This is preliminary,' people don't always pay enough attention to that," Aschwanden says. "By virtue of even covering it and making it a story you're giving it attention that it wouldn't otherwise have." If a study is so weak that it warrants only critical coverage, she says, "then it might be a good decision not to give it more attention with a story."

Reporting on preprints requires that journalists take special precautions to vet the research they present, says Camille Carlisle, science editor at *Sky & Telescope* magazine. She says journalists should check whether the preprint document indicates it is has been submitted to a journal. "If there's no comment indicating that the paper is in the process of being published, then remember that it might be because the results aren't worth publishing," she says. "There are also scientists for whom the arXiv is essentially a brainstorming blackboard. Proceed with skepticism."

That advice may hold especially true for preprints in the biomedical realm because of their potential consequences for public health, suggests Matt Davenport, a reporter at *Chemical & Engineering News.* Whereas he solicits comment from about three independent experts for stories about peer-reviewed papers and proceeds if he gets comment from one of them, Davenport's bar is higher for preprints. "For a preprint, maybe I double that and then follow up with folks who don't get back to me to find out exactly why," he says.

Likewise, Aschwanden says she seeks comment from multiple independent sources who can assess whether a preprint has a reasonable study design and can confirm that the experiment's goalposts were not moved during the course of the study. She also checks whether an unusually large number of participants dropped out of a clinical trial while it was in progress—and if so, why. Finally, Aschwanden recommends asking outside commenters to weigh in on the manuscript's statistical robustness. She notes that the American Statistical Association can be an excellent resource for science reporters.

It's also crucial to get context for a preprint—and when writing, to give readers a sense of the broader research landscape. That includes getting context about the scientists involved in the study. Anyone writing about preprints should "look at previous research the authors have published—both to check their credentials, and also to see how those findings may contribute to their latest work," says Melody Schreiber, a freelance journalist and editor of a forthcoming nonfiction book on premature birth called *What We Didn't Expect*.

Although scanning social media doesn't replace speaking directly with independent sources, doing so can provide an initial picture of how other scientists have responded to a new preprint. (Some preprint platforms, including bioRxiv and medRxiv, even display, on each manuscript's page, a rundown of how it has been mentioned by blogs and in tweets.) But Aschwanden cautions against relying on online discussion to guide decisions about whether and how to cover preprints. "You have to be careful because there are so many competing interests there," she says. When looking for outside comment on a paper, she says, "you want people who don't have a dog in the fight."

There's no magic number of independent sources to consult on preprints before proceeding with a story. Oftentimes, it depends on how big a finding the manuscript describes. "The higher the stakes of the paper, the more sources you contact," says Ed Yong, a staff writer at the *Atlantic*.

Sometimes, independent sources may have differing opinions when asked to comment about the value of a preprint. Reporters and editors have to decide how to handle such situations on a case-by-case basis, Morello says. In some cases, she says, one or two outside commenters may flag an important caveat to a study or even share serious reservations, but the study might still have news value for the general public and be worth covering.

Other times, those caveats might be deal-breakers. "Sometimes after hearing a bunch of folks say a study is good," Morello says, "you talk to somebody who brings up a statistical objection and then you ask other people about it and they say, 'Oh wait . . . that . . . eek!' And then you might not end up writing the story." Ultimately, covering contentious preprint studies requires journalists and their editors to make careful decisions on how to proceed—there's no easy rule of thumb.

Communicating Caveats

Just as important as the decision about whether to cover a preprint manuscript is how the research is described to readers, both in the story itself and on social media. Any coverage of a preprint should clearly convey that the study hasn't been vetted in the same way published papers have been. But Oransky cautions against simply saying that a study hasn't undergone "peer review" because lay readers might not be familiar with that phrase.

"As shorthand, it's better than nothing, but what I much prefer is to say no one has formally reviewed and critiqued this paper in [the] way a scientific journal might," says Oransky, who also teaches medical journalism at New York University. Similarly, Aschwanden recommends using straightforward, accessible language, such as by writing that the work "has not yet been checked for errors."

Even seemingly small wording choices make a difference. For ex-

New Tools Put Spotlight on Preprints

Recently, some groups have tried creating resources to help nonscientists gain more context about preprints. In 2018, Jonny Coates and a group of about 100 volunteer collaborators, all early-career scientists, established a highlight service called Prelights to spotlight interesting preprints in the life sciences. The group has now added coronavirus research to its purview. In April, they debuted a website in which COVID-19 preprints with major shortcomings, such as inadequate sample sizes, are flagged in yellow. (A limitation of the project, as Coates notes, is that the team does not yet include any epidemiologists or infectious disease specialists.)

There are other similar efforts underway. As Hannah Thomasy reported for *Undark*, a new system for curating preprints, called Outbreak Science Rapid PREreview, allows academics to review outbreak-related preprints. The project, which is funded by the British charitable foundation the Wellcome Trust, has collected more than 60 reviews.

ample, describing a preprint document as a "manuscript" rather than a "paper" and referring to it as having been "posted" rather than "published" online helps underscore, for readers, that the document is still in a preliminary stage.

The same caution should apply to headlines. Morello suggests that, as with stories about peer-reviewed but preliminary studies, media outlets should "be careful that the headline matches the caution of the story." For example, titles might include words like "preliminary analysis" and "suggests."

The most careful reporting can be undermined by an incautious tweet, so how stories are framed on social media is also important. As Morello observes, some people don't actually click on the stories they come across on social media. That may be why, as she's noticed, some reporters make a habit of promoting preprint stories not in a single Tweet, but in threads that provide context and note caveats that might otherwise be found only in the article text.

It can also be risky for reporters to tweet about preprints they haven't reported on themselves, Yong says. That's why he cautions journalists against tweeting about a preprint manuscript (or even a published paper) that they've found if they haven't yet done reporting on the work. "To do so is functionally equivalent to writing a story without talking to anyone, which I think we can all agree is a bad idea," he says. "There may have been a time when this practice was acceptable, and when journalists could use Twitter as a *means* of reporting—as a way of canvassing opinion, or testing ideas. But this is no longer that time." Yong says that the stakes have become too high and the public is more vulnerable than ever to poorly vetted claims.

Peer-Reviewed Studies Also Require Caution

Amid the concern about unvetted preprints getting undue media attention, it would be easy to overcorrect, treating peer-reviewed papers as if they are flawless. As freelance journalist Wudan Yan wrote in a recent *New York Times* article on media coverage of preprints, published journal articles also sometimes receive exaggerated coverage. And Coates suggests that regarding peer review as the gold standard may cause some journalists to let their guard down and not critique published journal articles as carefully as they should. For that reason, he says, "published papers are potentially more dangerous."

Penny Sarchet, news editor of *New Scientist*, holds a similar view. "We don't automatically treat preprints like they're that much lower in quality than peer-reviewed published papers," she says. "We don't believe peer-review to be a gold-standard guarantee that science holds up." As she notes, many articles published in scientific journals don't ultimately hold up when others try to replicate them. And, she says, there have also been "really shoddy" studies published on COVID-19 since the pandemic began.

How much manuscripts change as they go through the peer-review process remains uncertain. In March, Jeffrey Brainard reported in *Science* that one recent examination of 76 preprint papers, mostly in genetics and neuroscience, found that of the 56 manuscripts that were ultimately published, most underwent relatively few changes after peer review. But the study was small and is itself still only in preprint form.

Still, its findings mirror the experience of London-based geneticist and writer Adam Rutherford. Many of the papers he used in the research for one of his books were preprints about paleogenomics that had been posted on bioRxiv. "It occurred to me that all of them were subsequently published in mainstream journals, so I checked to see how much they had changed, and the answer was almost not at all," Rutherford says.

There's an important bias inherent in such analyses, though: They only consider preprints that eventually get published in peer-reviewed journals. But not every preprint makes it to publication. A 2019 analysis found that around two-thirds of preprints posted between 2013 and 2017 were later published. And some studies are even retracted while still in the preprint stage.

If a study does change considerably between the preprint stage and publication after peer review, Oransky says, news outlets should include "an update at the top to explain what changed." And if a study is retracted, he says, "I'd add an update at the top noting that, and saying why it was retracted."

Sarchet says the stakes are obviously very high at the moment for preprints, given the global concern about the current pandemic. "We're being a lot more careful with the COVID crisis because there's obviously much more potential there for people to take them [the preprints] very seriously and apply them to their lives," she says. But, she says, it's also important to share preliminary findings with readers if those discoveries are robust and intriguing, to give them a sense of the evolving nature of knowledge. Understanding how scientists "are exploring new ideas and testing those out" is an important part of covering science, she says. "That's why we've always covered preprints."

38 Getting the Most out of Scientific Conferences

Rodrigo Pérez Ortega

THE FIRST TIME Nick Mulcahy covered a scientific meeting, in 1998, he felt like a tiny boat bobbing in rough seas, with no compass on board. It was a cancer meeting, and the session titles were mostly indecipherable to him. Mulcahy couldn't make heads or tails out of the presentations. "I was completely flummoxed," he remembers. On top of being mystified by the scientific content, he was also clueless about the process of reporting from a meeting. He didn't know whether to sit in the front or the back of the room. He had no recorder. And his notes were a mess: He didn't talk to anyone or ask any questions—he just relied on comments from the podium.

After a few hours, in an act of self-preservation, Mulcahy found a pay phone and called his editor in Maryland. She laughed out loud when she heard how panicked Mulcahy was. But at that point, there wasn't much she could do to help. "My first meeting assignment produced nothing but a lot of anxiety," he recalls.

Now a senior journalist at *Medscape*, Mulcahy attends about half a dozen cancer meetings a year and writes an average of six to ten stories from each. "Reporting from a meeting is overwhelming, even for experienced reporters," he says. But with experience and practice, he has developed several strategies—from how he does prep research to how he chooses sessions to attend and takes advantage of gadgets and social media—that help ensure he'll come away with captivating stories without losing too much sleep. "The more you do it," says Mulcahy, "the better you get at it and the less overwhelming it gets."

Advance Work

Perhaps the biggest key to successful meeting coverage is to begin preparing well in advance of the event. There's no single best way of doing so—every scientific conference is different. "My strategy varies depending on the meeting and the technicality of the meeting," says Kelly Servick, a staff writer at *Science* who regularly covers biology, medicine, and biotechnology meetings.

Journalists who are experienced at meeting coverage agree that the first and most important step is to go over the program weeks in advance, scanning abstracts to learn who will be attending, what topics will be discussed, and what the breadth and depth of the subject matter will be. Only then can you pick sessions to attend and prioritize them. (Remember to check out both talks and poster sessions.) Wading through the program book can be "enough to make your head swim," Mulcahy says, but it's the only way to get oriented and plan successful coverage.

Don't stop at reading the program before the meeting begins, though. "Talk to as many people as you can," Mulcahy advises.

Servick agrees. For large meetings with lots of concurrent sessions, she often calls session chairs or speakers to narrow down her options and create a ranked list of one to three sessions for each time slot. For smaller meetings with fewer sessions, she says, she tends to do less prep work but may still interview the organizer ahead of time to get a sense of the meeting's overall aim.

Don't forget to use your editor as a resource. Especially if you are not familiar with the field at a conference you're attending, or if you're not sure what angles to cover, ask for specific direction.

Another key part of advance planning is contacting scientists you want to interview ahead of time, so they can make time in their schedule for you during the upcoming meeting. You can also ask fellow journalists who have more experience on the beat for guidance in advance of the meeting—though be aware that it's not fair to ask them to give away the story angles they're planning to pursue. Mulcahy also suggests that conference newbies study how veteran reporters cover meetings. "Go to their websites and read their stories, and copy their stylings," he says. "Imitation is the greatest form of flattery. If you do this long enough, you will develop a style of your own."

Should You Follow the Crowd?

Major meetings normally employ public information officers (PIOs) to put together a press program that includes news releases and sometimes briefings on the most important or newsworthy studies being presented at the meeting. "Good stories are more hit-and-miss than when a good PIO has made some advance information available to guide reporters," says Harvey Leifert, former public information manager for the American Geophysical Union and now a freelance journalist.

"If that is in place, the job is so much easier," Mulcahy agrees. A press program also gives you easy access to authors and outside experts without having to chase them after a panel. "It expedites the process," he says.

PIOs can help bridge "the journalistic and the academic world at meetings," says Ana Claudia Nepote González, communications coordinator at Escuela Nacional de Estudios Superiores Unidad Morelia in Mexico. In addition to organizing a formal press program, she notes, meeting PIOs can also help journalists locate and line up interviews with scientist sources, even if they're not included in the press program. After all, PIOs are the expert organizers of the meeting, so they know what's going on firsthand. Journalists can make use of their expertise by asking for suggestions based on their publications' needs and interests.

But it's smart to think of a press program simply as an aid, not a replacement for enterprising reporting. For one thing, organizations, program committees, and PIOs have biases and, in some cases, conflicts of interest that can color what goes into the press program. And PIOs may simply leave some of the best story ideas out of their press program. "My level of reliance on a press program depends on whether it contains actual breaking news briefings," says Servick. If a press program contains only "timely topic panels," she's more inclined to go find her own stories.

Some journalists steer clear of meetings' press programs altogether. "[PIOs] know what stories they want to tell, and they make it very easy for journalists" to tell those stories, says freelance journalist Aleszu Bajak. "It's a waste of time if you're chasing what the rest of the herd is chasing."

Best Seat in the House

Where you sit during meeting sessions makes a difference—and your first question shouldn't be "Where am I comfortable?" but rather "Where will I get the best reporting done?"

"I always sit in the front row," says Bajak. "I don't understand why journalists would sit in the back." Being in front is important if you want to take photos during the session—either of the panelists or of the slides—for future reference.

Planting yourself in front also means you can turn around and read the name tags of people who speak up during the Q&A portion of a session. Since the people who ask questions are often working in the same field as presenters, they can often be the best sources to comment on the presentation. Finally, being in the front of the room ensures you can be the first person to approach panelists when they come down from the podium at the end.

At the same time, bear in mind that some sessions might not yield good stories. If you suspect you might not stay for the entire session, you might decide to sit in the back or on the edge of a row in case you decide to abandon the session.

Taking Good Notes

Every journalist has different preferences for taking good notes at scientific meetings. Bajak uses a reporter's notebook; Mulcahy take notes in an unlined sketchbook. Bajak uses his iPhone's Voice Memos app to record talks; Mulcahy uses a digital recorder with an external microphone. Both Bajak and Mulcahy say that when they hear a nice quote, or when a speaker starts to talk about something important, they jot timestamps in their notebooks so they can go back and relisten to the relevant parts. "I don't even listen to most of the tape," says Mulcahy. "I just go to the time points and listen to those."

Other journalists prefer to use a laptop or a tablet instead of taking notes on paper. Servick takes notes on her laptop and records the talk with the Microsoft Word (for Mac) audio notes function, which records the audio in the room using the machine's built-in microphone and makes digital timestamps as the user types in notes (other programs that sync audio with typed text include Microsoft OneNote, Pear Note [available only for Mac and iOS], and Notability). If she needs to revisit something quickly, she can click that section of her notes and the program plays back the audio that was recorded when she was typing that material. "It's a pretty decent system," she says.

Servick also carries a Sony digital recorder, just in case her laptop

It's Okay to Jump Ship

There is no shame in leaving a session if it's too complicated or if the material seems dull or otherwise unlikely to engage your publication's audience. "You get to a certain point, where you're still processing what they were saying five minutes ago and they've already moved on," says Servick. "There's a certain hopelessness that ensues at that point, and I've definitely cut and run from sessions before, just because I thought my time would be better spent starting over and trying a different session." Servick says she typically decides to bail out on a session if:

- The subject matter turns out to be highly technical and without obvious broad implications, or "is generally putting me to sleep."
- She's so profoundly confused that she thinks it's unlikely she'll be able to catch up, even if she can snag the speaker for a few minutes after the talk.
- There's another session or talk about to start that sounds more promising.

dies. When she wants to ask a panelist a question after a talk is over, she uses a Livescribe smartpen, which records audio while she writes in a special notebook that converts handwritten notes to type and also syncs with audio. Later, she can see the notes on the computer and click to listen to the synchronized audio.

Leifert mostly takes notes on his iPad mini, using an external keyboard and the Notability app. He often revisits audio when reviewing his notes, slowing down the replay to type full quotes accurately, he says.

Other note-taking apps and programs that rank among journalists' favorites include Evernote, Penultimate, Google Keep, and iOS Notes, all of which sync notes between digital platforms, so you can take notes on one device and revisit them on another. Another option is Cogi, a clever phone app that keeps audio in a buffer; when you hear something important, you can tap the screen and Cogi backs up to record the last 15 seconds and keeps going until you tap the screen again.

Meeting Coverage by Hashtag

Social media, and Twitter in particular, is also a powerful tool when covering conferences with several concurrent sessions. You can't be everywhere at once, but Twitter allows you to follow what's going on—and see whether the grass is greener—in other sessions.

Most meetings—and sometimes even individual sessions—have an official hashtag, so looking for quotes, session photos, and potential story angles is easier than ever. (It should go without saying that if you quote a researcher based on someone else's Twitter feed, it's essential to independently verify that the quote is accurate.)

With so much information being presented, "Twitter helps you find the needles in the haystack," says Mulcahy. "What people are really doing is putting the spotlight on things for you." Twitter can also act like "a public, virtual notebook, where I can refer to my own tweets," says Bajak.

More Than News

Covering a scientific meeting can be an opportunity not just to report on new results, but also to build your network and get scientists to trust you, says Nepote González. Meetings are also a kind of early warning system for studies coming down the pike. Some scientific studies presented at conferences are still in progress, and learning about them at that stage gives you a chance to connect with the scientists doing the work, understand the context for the research, and ask to be alerted when the completed study is accepted for publication.

Besides being a source of scientific results, meetings can also offer glimpses into other aspects of the scientific life. Mulcahy, for example, recalls a talk at an oncology meeting during which a prominent doctor unexpectedly revealed his HIV status and the stigma he had faced. For Mulcahy, the moment lent unexpected personal resonance to a story that might otherwise have felt entirely academic. "Always be looking for a story," Mulcahy says. "We are more than just data reporters. Of course, sometimes that's all we have time for, but the job is a lot more interesting if you can find the politics or the human drama at hand. And it is always there."

Other important and telling events can happen outside the lecture halls. At another meeting Mulcahy attended, an activist group of patients protested outside the meeting venue, and he was the only journalist that covered it.

At most meetings there is also a social program that includes receptions, award ceremonies, dinners, and parties. "They don't usually generate straight news stories, but they can lead to more casual and frank conversations, and connect me with new sources I can follow up with later on," says Servick. At medical meetings, Mulcahy also tries to engage with patients who attend such events. "It gives meaning and heart to all of the technical terms," he says.

39 Interrogating Data: A Science Writer's Guide to Data Journalism

Betsy Ladyzhets

IN MANY WAYS, all science writers are already data journalists. What do science writers do when they report on a newly published study? They dive into the details of the paper's results; they ask experts for opinions on potential flaws in the methodology; they seek to connect the conclusion to their readers' lives. Such investigation is driven by a desire to find evidence from the most authoritative sources and present it as clearly as possible. Reporting on data requires the same skill set.

Both data journalism and science writing boil down to "taking something complicated and trying to make it understandable," explains Sara Chodosh, an assistant editor and graphics producer at *Popular Science*. Once a data journalist has answered their own questions, they go through the same process with an imagined reader: What can the data tell the audience that will help them grasp a larger pattern or concept?

Data journalism, whether it takes the form of a static visualization, an interactive feature, or simply a bit of additional analysis to add context to a breaking news piece, can bring scientific results to the foreground of a story and make them accessible for readers. Take, for example, the maps in a Reuters article that put Australia's bushfires into perspective. Marvel at an interactive astronomy chart published by *National Geographic* that allows readers to explore our solar system's moons. Explore a *Climate Central* report on shifting snowfall levels, which invites local journalists and meteorologists to repurpose the data in order to connect changing weather patterns directly to their audiences.

In its simplest definition, data journalism is the practice of using numbers and trends to tell a story. It requires a variety of skills: research to find the correct dataset, analysis to determine what kind of story this dataset may tell, and presentation to share that story with readers. And these skills are within reach for many science writers, even without any programming background: Simply ask questions, and you will find the central tenet of a story.

Research: Choose Your Data

The first step with any data story is finding a dataset to analyze. For science writers, one natural source is the results section of any paper that you believe tells a compelling story. Many scientists release their unanalyzed data on open-access platforms such as Dryad and GitHub, a practice that allows others, whether scientists or journalists, to explore and build upon published results. And even data that are not shared through open-access channels are often available on request.

Either way, the choice to use the results of one particular study in a data story requires careful vetting; consider the authors' credentials and pore over their methods section before diving in.

Priyanka Runwal, a science writer and data reporter at *Climate Central*, points out that the process of finding a dataset may depend on the assignment. In some instances, one may have a question in mind (say, "How many Americans have been tested for COVID-19?") and search for a specific dataset that answers this question. In others, one may come upon an intriguing dataset (say, the Global Health Security Index) and seek to formulate a question from it.

In examining a potential dataset for use in a project, consider whether the data tell a compelling story. Are there evident trends or interesting outliers? Would readers want to explore a figure, or would they prefer to jump ahead to the conclusion? A story explaining a review of biodiversity hotspots, for example, may benefit from a map or chart showing where these habitats are located around the world and how they are threatened by humans. In contrast, focusing heavily on numerical results from different trials in a story about testing for a new medical treatment may distract readers from understanding the qualitative conclusions about what the treatment so far seems to accomplish and the necessary steps to come.

Besides these questions of reader value, consider logistical concerns. Are the data downloadable? Have they been released under Creative Commons licenses? What do all of the data labels represent? Do you understand the study methods, caveats, and implications, or will you need to ask a scientist or press officer for clarification?

Analysis: Rely on Your Curiosity to Turn a Spreadsheet into a Story

Once you have a dataset, the next step is to find patterns in the numbers. Data analysis can often feel like chipping away at a stone in order to make a sculpture; you may start with a massive spreadsheet and spend days isolating specific variables or data points which will illustrate a trend to your readers.

You may make this process more targeted by asking questions of your dataset as though it is an interview subject, suggests Peter Aldhous, a science reporter at *BuzzFeed News* and data journalism professor at the University of California, Santa Cruz, and the University of California, Berkeley. As he says: "What can the data tell me that I want to know?"

Common questions to consider may include: How do you need to clean the data (through standardizing names, changing labels, geocoding, and so on) to ensure that categories match up and all necessary information is present? What role does each variable play in the source study or in other similar datasets? Which variable may be used as an indicator of a larger trend? What analysis is necessary to show that trend— for example, what other variable might you compare to the first, or what groups of data points might you compare to each other?

Don't let your curiosity run too wild, though. Aldhous cautions that, like other sources, data can mislead you if you aren't careful.

Duncan Geere, a freelance data journalist and former editor at Information Is Beautiful, puts his warning this way: "Figure out what the data is showing, but also what it's not showing." What are the limitations in this dataset, due either to flaws in the methods used to compile the data or to discrepancies between what the data reveal and the story you want to tell? How might you want to filter the data to account for limitations, outliers, or missing pieces? What biases may have been present in the compilation? Closely examining data-collection methods is especially crucial when the data are describing people.

Geere suggests writing down aspects of a dataset that you find interesting, as well as questions that come up, as you explore the data. "I reason that, if I find this particular aspect of the data interesting, then my audience will as well," he says.

It may take some time to home in on what variable or trend from a dataset tells the most compelling story. Runwal leans into this exploration, she says. "For me, it requires patience, and eyeballing numbers for a while to actually make sense of them." To this end, you may test several different methods of filtering or analyzing your data before deciding which focus will be most informative for your readers.

Patience is also key in the analysis process because code (even a supposedly simple Excel formula) often breaks. When that happens, online resources abound: forums such as the National Institute for Computer-Assisted Reporting (NICAR) listserv, Stack Overflow, and even social media sites can help you solicit advice from more experienced data reporters. Geere recommends the Data Visualization Society, which boasts an active Slack server including both journalists and visualization experts from other fields.

Finally, just as responsible writers record interviews and save their notes, responsible data journalists keep careful track of every step in their analysis. You want your work to be reproducible, both by other people in your newsroom—data journalists aren't safe from copyediting and fact-checking—and by readers.

As Sam Leon, data-investigations lead at the international NGO Global Witness, explains in a chapter of the *Data Journalism Handbook* on methodologies, data can easily be "distorted and mis-represented" through errors at analysis stages. Such errors can range from a typo introduced while cleaning data to an analytic choice that misrepresents correlation as causation.

Resources for Data Journalism Novices

Several free resources can help you build data visualizations without coding. (But fair warning: If you travel down the path into the world of data reporting, you may find yourself seeing coding as a means of accomplishing more complex and more customizable presentations.) For more coding resources, check out NICAR, as well as data journalism courses on Coursera, Code Academy, and the Northeastern University School of Journalism's *Storybench* publication. For a list of journalist-friendly data sources and popular programs that data journalists use for analysis and to build data visualizations, visit bit.ly/TONdatajourn.

Presentation: What Should Readers Take Away from a Data Story?

Just as your questions can drive your data analysis, potential questions from your audience can drive your presentation. "Good science communication thinks about its audience," Geere says. Good data visualizations do the same; they tell a story that the audience will be able to follow, whether that audience is highly science-literate readers of a trade publication or young readers of an educational site.

Geere outlines the basics of storytelling through data in a blog post: Like any other story, he explains, visualizations need a beginning (an entry point), a middle (answers to readers' key questions), and an end (a final takeaway for the reader, whether this is a better understanding of a scientific issue or a connection to their own life).

Different visualization formats can highlight different aspects of a dataset. Kaiser Fung, data-science expert and founder of the blog *Junk Charts* (which highlights errors in data visualization in the media), lays out some ground rules in an article for the *Data Journalism Handbook* site. Pie charts (if they can't be avoided) should be designed with careful consideration to color and order of sections, as readers' eyes will be drawn to the largest sections. Bar charts and dot plots allow for easy comparison between groups. Scatter plots call attention to trends, and regression lines may be added to guide readers' interpretation.

But there are more ways to present data than in static charts. In recent years, data reporters have increasingly sought out new ways of making their stories interactive, from Johns Hopkins's COVID-19 tracker, which shows the virus's global spread, to *Stacker*'s data-based slideshows, which add photos and context to each figure in the datasets upon which they rely. (Disclosure: *Stacker* is my employer.) Interactive features can help readers narrow or broaden the scope of a story according to their interest, to see how the data directly apply to them. And such features do not necessarily require extensive coding, either; searchable databases, for example, which are essentially public spreadsheets hosted by journalistic organizations, are a useful tool for readers to find specific information and do their own research.

However you present your data work, though, one guideline is always relevant: Make it simple. "I am always striving to make things that you can look at and immediately, as clearly as possible, understand what you're being shown," Aldhous says.

Chodosh agrees, noting that the more data are pared down, the easier it is to follow a story. Ensure that readers can follow one variable or one group of values at a time, and test your visualization by showing it to colleagues who aren't familiar with the data. Simple color schemes, large text, and clear captions can also help make visualizations more accessible to readers who may otherwise have trouble following them.

In addition to considering your audience in the presentation of your data itself, consider your audience in writing a methodology section. A methodology can be a direct link to your code, a precise series of steps, or simply a paragraph at the end of your article. The complexity and location of your methodology section should depend on your audience: How much do you anticipate that this audience will want to understand precisely how you arrived at your conclusions?

In its most basic form, a methodology section should include a clear link to your data source and the major steps you took to analyze the data, written in simple language without jargon, as well as any caveats or major exceptions.

40 **Explaining Complexity**

Carl Zimmer

IMAGINE YOU'RE A crime reporter writing a story about a shooting at a nightclub. Now imagine that none of your readers know what a gun is.

Suddenly, your story got a whole lot harder to write.

You can't just jump into the shooter's backstory, or the victim's suffering, or the detective work that led to an arrest. Instead, you've got to explain the chemical properties of gunpowder, the physics involved in an explosion pushing a bullet down a gun barrel, the speed at which a bullet strikes a victim, and the effect that such a projectile has on the human body. And as you work on your explanation, you start wondering about other things you may need to explain. Maybe you need to explain how bullets are made, or why the materials in a bullet make it so deadly. Maybe you should explain the mechanism by which the shooter's Beretta automatically loaded a new bullet in the chamber after each shot, so that your readers will understand why the victim ended up getting hit by so many bullets in so little time. The possibilities become paralyzing.

Welcome to the science writer's dilemma. Science writers tell stories about things that many readers are unfamiliar with. Things like quasars, bosons, reversals of the Earth's magnetic field, and the mating habits of bedbugs. Science writers can't simply write stories about these things. They have to explain them along the way.

A good explanation achieves a happy medium between too little and too much. If you assume that your reader knows as much as you do, you will be prone to leaving out crucial information. It can be hard to notice what's missing from an explanation, because every part of it exists in your mind, if not on the page. You read your drafts in the same way we look at optical illusions and fill in the blanks to create a complete shape.

There are only two ways to avoid this mistake: either have someone else read your story—someone who's not an expert on the subject, of course—or develop the ability to override your own in-filling instincts. Paradoxically, the more expertise you have in a subject, the harder this journalistic brain training will become. If you learned about superstring theory 20 years ago in grad school, it will be challenging for you to imagine what it's like to be someone for whom superstrings are not as simple to understand as parallel parking (or perhaps easier, depending on your driving skills).

But filling in the gaps is not the same as burying your reader alive. It's a mistake to assume that in order to explain something, you have to deliver a semester-long introductory lecture course. It's true that we can all learn a lot from a semester-long introductory lecture course, but we don't expect to enroll in one whenever we open up a magazine or visit a news website.

What we expect, instead, are other experiences: a story, in some cases; an argument, in others. If you spend all your time explaining rather than telling a story or advancing an argument, the structure of your writing will collapse under that explanatory weight.

Thus, the most important step in explaining something well is to figure out what's the minimum amount of explanation required for readers to understand your overall piece. How little explaining can you get away with? Once you've worked that out, then you have given yourself a clear set of goals to achieve. You can then try to make your explanation as delightful to read as the most unexpected plot twist.

There are many tricks and tools you can use to build up good explanations. Metaphors, of course, are essential. Rather than trying to break down an explanation into a buzzing swarm of details, think of an image that neatly wraps up those images into one concept and conveys the gist of what you want to say. If you're interviewing scientists as part of your research for an explanation, ask them if there's a metaphor that they like to use. You'll be surprised how often people who deliver semester-long lecture courses can distill an idea into a few evocative words.

In some cases you deal with an explanation quickly and then move on. But sometimes you have to deal with some truly massive idea. In these cases, it's often a good move to disperse pieces of the explanation throughout your story. The only reason you can explain global warming to your readers, for example, is that for a couple centuries, people have been struggling to piece that explanation together. Tell the story of the explanation, rather than giving the explanation itself. While few people are ready to sit down for a semester-long lecture course at the drop of a hat, we're all up for a good story.

41 **How to Do a Close Read**

Siri Carpenter

PART OF SHARPENING one's craft as a reporter and writer involves understanding what makes notable nonfiction stories tick. One way to do that is to closely read stories that you've admired (or for that matter, stories that have irked you) or that have racked up awards, been included in anthologies, or otherwise gotten attention (including being highlighted in *The Open Notebook*'s writer interviews or Storygram annotations). But reading a nonfiction story from a journalistic perspective is about more than just comprehending the subject matter and noting whether you liked or didn't like the story. Examining stories more closely, for the purpose of studying the craft, usually requires reading them several times with some purposeful questions in mind, then giving yourself the gift of time to think about how you might incorporate some of your observations into your own work.

Below are some of the kinds of questions you can ask about a feature story to deepen your perceptions of it and to get you thinking about how writers make the most of their craft. These questions encourage thinking about stories on multiple levels.

Looking for a story to start with? You could begin with some stories *TON* has showcased through our interviews and annotations, such as Maggie Koerth's 2017 *FiveThirtyEight* story about Pan Pan, the panda who was "really, really, ridiculously good at sex." Or with Kathryn Schulz's much-acclaimed 2015 *New Yorker* story "The Really Big One," about the likelihood of an eventual massive earthquake in the Pacific Northwest's Cascadia subduction zone. Or with *Washington Post* reporter Brian Vastag's 2012 profile of dinosaur footprint tracker Ray Stanford.

A note of caution: Don't torture yourself trying to answer every question below while doing a close read. Instead, pick a manageable number—perhaps one or two from each category, for starters. If you do close reads regularly, it will become easier to pick up on important elements in the stories you read.

Your First Take

- Did you enjoy reading the story? Did you find yourself wanting to keep reading, to find out what happens next? Or did you find yourself drifting or have to force yourself to finish?
- What aspects of the story appealed to you? (The tone? Voice? Character development? A conflict or tension at the center of the story?)
- Did you want to tell others about the story? In a sentence, what would you want to tell them?
- Did you want to learn more?
- Did you wish you'd thought of the story yourself? If so, what might have led you toward such a story?

Story Topic

- Is the topic new or controversial? Why or why not?
- Has this topic received a lot of media attention? Why or why not?
- Is the topic of interest to you? Why or why not?
- Who might be affected by this issue?
- What are the political/economic/social ramifications of the topic discussed in this story?
- How should this story influence how we think about the subject matter?
- How does the story add or contribute to what's already been written on this topic or to public discussion on the topic?

Story Focus

- What is the story's central question or argument? Can you boil the central point down to a sentence? (Try the six-word test: Can you boil it down to just six words? Can you distill the story's essence down to a headline that communicates why readers should read it?)
- Does the most compelling information in the story support this central theme or focus? If so, how?
- Is there anything that, if handled differently, might have distracted from the main focus?

- Are there any disconnects or points that don't flow logically?
- Where is the story's main idea first articulated? How much detail is provided there?
- Are there any secondary themes in the story? What are they? How do they connect with the main theme?
- Consider the difference between topic and story. What makes this piece a story?

Information

- What are the most essential pieces of information that this story conveys?
- Does all the information belong in the story? Or is there any information that you feel should have been left out or that adds "color" for color's sake?
- What information do you think the writer left out? Do you think the writer left that information out unintentionally, or intentionally, so that readers can draw their own conclusions? Did you find yourself wishing the story had included some other information? Did the story leave any of your questions unanswered?
- What facts, numbers, statistics, quotes, anecdotes, characters, or details stand out? What makes them stand out?
- Does any information in the story show deep reporting on the writer's part? If so, how so?
- Does the writer use similes, metaphors, or analogies to explain concepts or processes? What are some examples, and why do they work well (or not)?
- Does the writer use particulars as stand-ins for something larger—a synecdoche?
- Does the writer convey some information (say, about a character or a situation) subtly, rather than directly?
- Does the story include conflicting points of view? How does the writer present these? From the way the story is written, is the writer's own opinion evident? Why or why not?
- If the story focuses on a particular approach or solution to a scien-

tific or social problem, does it present evidence of the real-world impact of that approach? Does it present limitations of that approach?

- Does the story contain speculation? If so, is there good justification for the speculation? How does the writer make clear that it is speculation?
- What insights did reading this story give you about the subject matter?

Structure

- What makes the story's lede successful (or not), in your opinion?
- Is the main organization thematic or chronological? Why did the writer choose a particular episode to begin the story? Did it succeed? How did the story move forward in time? Did it jump back-and-forth in time? Can you think of other ways of beginning and ending the story?
- Is the story organized into acts (like a play)? If so, what are the acts?
- Is there anything unusual or striking about the story's structure?
- Does the story have a nut graf (also known as a billboard)? How long is it? What does it convey? What does it leave out? Does it give away too much detail too early, or just enough to signpost where the story is headed?
- If the story does not have a nut graf, per se, how does the writer signal the story's main themes to readers early in the story?
- What kinds of transitions does the writer use to move readers from one idea to another? Do the transitions flow smoothly, or are they jarring? Are there any tricks the writer has used to make those transitions less obvious?
- Does the story have characters it comes back to and steps away from at crucial points, or does it stay with a central narrative line?
- Are there flashbacks? Where? What do they accomplish for the reader?
- What type of ending, or kicker, does the story have? Does it end on a quote? If so, whose voice gets the last word, and what does that convey? Does it summarize main points in the story? Does it give clo-

sure? Does it relate in any way to the lede? Does it point the reader toward the future in some way?

- Do you agree with the way the ending is written? Would you have done it differently? Why or why not?

Narrative

- What elements of narrative are present in the story? (Scenes? Dialogue? Character development? Conflict? Action? Sensory details?)
- What is the central conflict or source of tension within the story? Is that conflict resolved by the end? How?
- How does the writer create suspense and tug the narrative forward?
- Is there a main character in the story? Who is it? Or is the main narrative thread an idea or problem?
- How does the writer bring characters to life in the story? What actions and other details are key? In painting a picture of the main character, does the writer include any characteristics that readers might perceive as less than positive?
- Do the length and pacing of quotes add to your sense about a character's inner life?
- Is there foreshadowing? Where? What effect does that have?
- Is the writing cinematic? Does it pull back to a wider angle in certain "scenes" and zoom in for others?
- Is there only a single narrative thread through the story, or are there multiple narrative threads? If there is more than one, how does the writer weave them together?
- How is expository material, such as technical or historical background, woven into the narrative? How does the writer weave expository material into the narrative without slowing the pacing of the story?
- Is the writer themself a part of the narrative? Is that necessary? Effective?
- Choose a particularly evocative scene and reverse-engineer it: Does the writer include a physical description of any of the characters? Are the descriptions effective? If there's no physical description,

does that detract from the scene? How many senses does the writer highlight? What physical details does the writer include and, crucially, what details do they omit? How does their choice of details contribute to the scene's strengths?

Language

- What vivid nouns and active verbs give the story energy?
- Is the language in the story more formal or informal? What are some examples?
- What is the story's tone? Is it witty? Somber? Celebratory? Informative? Satirical? Chatty? Forceful? Optimistic?
- What types of language usage contribute to an accessible, conversational tone?
- Does the story contain technical language? How does the writer avoid confusing readers with technical or specialized terms?
- Does the writer use other devices, such as alliteration, assonance, or onomatopoeia to make the writing sing?
- When does the writer use adjectives and adverbs?
- Are there examples of parallel construction that make the writing more compelling?
- How long are the sentences? How variable are they in length? What effect does the writer's decisions about sentence length have?
- How long are the paragraphs? How variable are they in length?
- Does the writer ever intentionally violate conventions of grammar or usage? Where? What effect does that have?
- Would you describe the writer as distant from or close to the reader? What effect does that have?
- What kinds of words does the writer place at the ends of sentences and paragraphs? What effect does that have?
- What senses does the story draw on?
- Does the tense change at any point? Why might the writer have done that? What effect does it have?
- Does the voice change from active to passive, or vice versa, at any point? What effect does that have?
- Does the writer use humor or satire to affect the story's tone?

- Does the writer use any literary or pop-cultural allusions? What effect does doing so have?
- Are there moments when the writer changes voice or perspective to match the tone of one or more characters? This is sometimes called "writing in character" and can effectively communicate a subject's personality.

Art and Multimedia Elements

- Does the story contain memorable visual elements, such as photos, illustrations, charts, graphs, maps, or videos? If so, what makes those elements memorable?
- Do the visual elements of the story merely "decorate" the piece? Or do they substantially contribute to readers' understanding of the topic or narrative? How so?
- How does the lead art in the story convey important themes or draw readers into the story?
- If the story contains portraits of people, do they help communicate something meaningful about those characters? If so, how?
- What emotions do the art or multimedia elements in the story evoke? Is that appropriate to the piece? Why or why not?
- If information is presented through data visualizations, does this presentation make the information easier to understand? Is there any aspect of the presentation that might be misleading?
- What information is left out of visualizations in the story? Why might that information have been left out?
- How does the style of visualizations contribute to their value?
- Are text and multimedia elements presented in a way that makes it clear how best to navigate the story? Or is it easy to lose your place in moving back and forth between elements?
- Is the placement of multimedia elements important to their impact?
- Do the multimedia elements just repeat what's in the written story, or do they expand or enrich the written story in some substantive way? Do they support the written story, or distract from it?
- What multimedia elements could further enrich the story, if they aren't already present in the story?

Take Your Close Read to the Next Level

- What are the three most important things the writer did to make this story so good?
- What are three things that can elevate a fairly good story into an excellent one?
- What are three key observations you made in the course of close-reading this story that you'd like to remember for your future work?
- What are three concrete things you can do to help enhance your own reporting or writing, based on observations you've made here?

42 A Conversation with Maggie Koerth on "The Complicated Legacy of a Panda Who Was Really Good at Sex"

Ed Yong

TODAY, THERE ARE 520 pandas living in captivity, and fully a quarter of them are the descendants of an especially fecund and recently deceased individual named Pan Pan. Like an ursine cross between Don Juan and Jean Valjean (he is simply Panda 308 in the official studbook), Pan Pan seeded the zoos and research centers of the world with his progeny. You merely have to gesture at his family tree, with its multiply sprouting branches, to counter the inaccurate stereotype that pandas are bad at sex. As Maggie Koerth wrote in a 2017 feature published at *FiveThirtyEight*, "Pan Pan saved his species by being really, really, ridiculously good at sex."

But in her masterful piece, Koerth also shows that Pan Pan's story is a complicated one—as much about conservation's limitations as its triumphs. It's about animals that become too used to captivity and breeding programs that become almost too successful for their own good. It's about the unpredictable consequences of well-intentioned plans. And it's about our inability to fully heal the wounds we inflict on the world. In Pan Pan, Koerth found not only a protagonist but a black-and-white avatar for an issue that is all shades of gray.

"The Complicated Legacy of a Panda Who Was Really Good at Sex" was published on November 28, 2017. In this interview, Koerth talks about the making of the piece.

ED YONG: I remember talking to you about this story in late 2016. Over a year later, it's finally out. Can you remember a time when you weren't constantly thinking about—and I mean this quite literally—a fucking panda?

MAGGIE KOERTH: I don't think my husband can. This started in August 2016. There was a panda about to give birth and I thought: There's probably all this data on panda sex, since it's so heavily managed. I would get the data, make a little chart, and get this done this week. My editor, Chadwick Matlin, had this great idea that we were going to do more short-turnaround data-analysis stories—a chart and 400 words. And they kept on turning into 1,500-word stories by the time you framed the nuance around the data.

So, did you email people cold, saying, Hey, is there a list of all the pandas who've had sex?

Yes, that's basically what I did. The first emails I got back just said, The data exist. Then a guy called Henry Nicholls wrote a book about pandas and sent me a Word doc of the giant panda studbook from 2006. Around the same period, Ron Swaisgood from the San Diego Zoo sent me a printout from 2013. Neither of them would convert into easily searchable Excel files. So that was fun.

Is that why the story took so long?

Partly that, but partly because I was also trying to figure out what the story was. For a while, it was an animal welfare exposé. One of the first people I contacted was Kati Loeffler, who used to be director of animal health at the Chengdu Research Base of Giant Panda Breeding. She had a lot of things to say about animal welfare in Chengdu, but after a few months, it became apparent that I wouldn't be able to verify some of her claims. Some were around what she saw as profit motives for the Chinese government. That's beyond my pay grade.

Because you're probably not going to FOIA China about pandas.

Right, although I did FOIA the US over the panda contract. [This is the document that says all pandas in American zoos are loans from China, and that the US pays them a million dollars per pair per year, plus expenses.]

In the meantime, I was able to get some of the data in the Nicholls document analyzed, and I saw that one panda—308—had way more offspring than anyone else. I mentioned that to Jonathan Ballou at the Smithsonian and he said, Oh, that's Pan Pan! It was the first time I really understood that he was as important as he was. In February 2017, I thought, Okay, this is a biography of Pan Pan. This is not just a story about pandas, it's a story about this one panda. One of the things I read around that time is this kind of classic profile of Frank Sinatra . . .

OMG. The next question I was going to ask was, Is this "Pan-Pan Has a Cold"?

Yes, this is "Frank Sinatra Has a Cold," but about a bear.

"Frank Sinatra Has a Cold" is this classic New Journalism–era profile feature, written by Gay Talese for *Esquire* in 1966. The basic conceit is that Talese gets this assignment, but then can't get an actual interview with Sinatra. So, instead, he kind of ends up interviewing all these people connected to Sinatra, some heavily and some kind of tangentially. I started thinking about it—and went back and reread it for the first time since journalism school—when I decided that Pan Pan was my way into this story, that he was the main character. I was having a hard time because I really know jack shit about Pan Pan. Because he's a fucking bear.

One of the things my sources kept talking about was their frustration with the way pop culture anthropomorphizes pandas. We place these expectations on them. We give them interior lives we can't possibly know. It's like the cult of celebrity. So I thought, I don't know Pan Pan's interior life, or even all these details of his biography. But I *can* take the Gay Talese route—talk to people who crossed paths with him, people he meant something to. And tell that second-hand/third-hand narrative, where we are explicitly doing the thing we normally do only implicitly, which is tell ourselves stories about these animals.

I had to go back to all of these sources who I had talked with for hours about animal welfare and breeding studbooks, and ask them about Pan Pan. Why do you think he's important? What does it mean to have such a big swath of this captive population related to Pan Pan? I also started doing Google Scholar searches for specific research papers that mentioned him, and I came up with [a] 2004 paper on panda mating success factors. Something about that paper made me feel like I was profiling a real, living thing that had a personality and had once taken up space in the world.

I also found this woman from Pandas International who was a giant Pan Pan fan and had gone out to meet him several times. And I found this entire online panda-fan community. There are some people who are *really* serious about it, in a fan-fiction way.

Pan fiction.

Effectively. They'll make videos and write stories and talk about the individual pandas like they're their babies. I found Chinese news articles about his history through those [online] groups.

It feels like you went full circle, from thinking that this is a story about pandas in general, to figuring out that it's a biography of one panda, to realizing that the one panda is a symbol for conservation—and its limitations. As you write: "Once you break a species, you can't easily put it back together again." How did that last bit crystallize?

Some of the things that fed into it came up completely accidentally in early interviews. I talked to a guy who had written a paper on the limits of breeding programs. I learned about condors and Florida panthers. [California condors were saved through captive breeding programs, but can't survive on their own because their habitat is still full of toxic lead shot; Florida panthers became so inbred that their population had to be supplemented with diverse females, airlifted in from Texas.] I repeatedly read through these interviews and connected the dots.

The broader complication is that captivity isn't necessarily a great place to be, and that once you go into captivity, you can't go back. I wanted to get that across. Pandakind is doing really well, but to get there, maybe Pan Pan himself didn't live his best life. How do we feel about that? It became really important that this dark pivot happened.

And yet, the piece also has a distinctive lightness of touch even when dealing with dark themes. This, for example, when describing breeding programs that select pairs with the goal of eventually releasing offspring into the wild: "Big, beautiful, independent bear, seeking same. Must not love humans." That's partly why, as with many of your stories, I can tell that it's yours without having to see the byline. Early-career writers often think about finding their "voice"; how did you find and keep yours?

I spent a good part of my early years freelancing absorbing the voices of publications and vomiting that back out onto pitch letters. I still feel that's insanely good when you're starting out because it makes you [seem to editors] like you need less editing. I don't know where that stopped and I started sounding like myself. Some of it is just humor. A lot of it is writing the way I would tell someone about this if I was sitting around at a bar. If somebody reads this and walks out the door and goes to a dinner party, what are the two or three things I want them to say? I fashion the rest of the story around those *whoa* moments. That has made it easier to write long stories, and also to find my voice in them.

After the panda story was published, I had a panic moment. Oh my god, did I just write a *Mental Floss* story? A light sex-joke romp? And no, I thought, you're a grown-up and you've figured out how to have emotion in the light sex-joke romp. That's part of the process of finding your voice, too. Learning how to be funny eventually enabled me to be deep with the funny.

But you don't want it to be all humor. You want that dark pivot, but you want to make sure that's telegraphed in such a way that it doesn't seem to come out of left field and feel jarring. You want to make people laugh and make them feel uncomfortable or disquieted and make them feel pathos all in the same piece.

And did you ever get the hang of doing short data-driven stories?

No. I think the panda story put the final nail in that coffin.

Acknowledgments

I'M TREMENDOUSLY GRATEFUL to have the opportunity to publish a second, significantly expanded version of this book. I'm grateful to all who helped make that possible.

Cat Warren, a friend and supporter of *The Open Notebook*, introduced me to her spectacular agent, Gillian MacKenzie. Gillian was an immediate champion for this book, and helped me shape the vision for how we could expand it and bring it to new audiences. Joe Calamia of the University of Chicago Press has been an insightful and enthusiastic editor, and I'm thrilled and honored that he saw promise in a partnership with *The Open Notebook*. Thanks also to the rest of the team at the Press who made publication of this second edition possible, including Carrie Olivia Adams, Skye Agnew, Matt Lang, Annie Leue, and Joel Score, and to June Sawyers for the index.

A number of people have provided editorial and technical assistance, as well as moral support, for the preparation of the second edition, including especially my dear friends Jeanne Erdmann, Julie Rehmeyer, and Alexandra Witze; Sarah Luft, a much-valued member of the *TON* editorial team; my daughter Alia Carpenter; and my husband Joe Carpenter.

I remain eternally grateful to all who supported the first edition of this book, when it was a total gamble. In particular I'm grateful to Russ Campbell of the Burroughs Wellcome Fund, whose support for the first edition is the gift that keeps on giving. I'm also indebted to the journalists who kindly contributed wise words in the section introductions to the first edition: Azeen Ghorayshi, Christie Aschwanden, Maggie Koerth, Apoorva Mandavilli, and Dan Fagin. Though revising and expanding the book made it necessary to omit those pieces in the second edition, I'm no less grateful for their generous insights.

I also would like to thank the members of *The Open Notebook*'s editorial team—Saugat Bolakhe, Torie Bosch, Aaron Brooks, Inés Gutiérrez

Jaber, Emily Laber-Warren, Sarah Luft, Rodrigo Pérez Ortega, Debbie Ponchner, Sandeep Ravindran, Jill Sakai, Kelly Tyrrell, Katherine J. Wu, Rachel Zamzow—and of our board of directors—Jeanne Erdmann, Jane C. Hu, Alexandra Witze, Shraddha Chakradhar, Ann Finkbeiner, María Paula Rubiano A., Ashley Smart, and Sisi Wei. This book, like all the work we do at *TON*, would not be possible without their support and dedication to our mission.

Acknowledgments to the First Edition

This book owes its existence to the talents and goodwill of many people. I'm grateful, first, for the almost three dozen journalists who contributed articles. Their insights, sensibilities, and collective wisdom animate every page of this book. My deep thanks also to the scores of journalists and others who shared their time, experiences, and knowledge within those articles.

This book would also not have been possible without Jeanne Erdmann. It grew out of the nonprofit journalism website *The Open Notebook*, which she and I launched a decade ago. Over the years, Jeanne has brought countless creative ideas to *TON*, including the popular Pitch Database, the *Ask TON* advice column, and the Natural Habitat series. I am deeply grateful for her unflagging enthusiasm for what started as a labor of love, and for her friendship.

This book was made possible by a generous grant from the Burroughs Wellcome Fund. BWF's Russ Campbell has been a relentless champion of *The Open Notebook* for many years, supporting our early-career fellowship program and other endeavors, and providing essential support for *The Open Notebook*'s growth.

TON has also been fortunate to have support from Science Sandbox, an initiative of the Simons Foundation; from the Council for the Advancement of Science Writing; from the Gordon and Betty Moore Foundation; from the Therese Foundation; and, in our earlier days, from the National Association of Science Writers and the Knight Science Journalism Program at MIT. We're also thankful for the individual donors whose gifts are essential to our work.

I began planning this book at a writers' master class in Bethany Beach,

Delaware, led by the great Jacqui Banaszynski. She and the other writers at what I think of as Camp Jacqui—Jill Adams, Lauren Gravitz, Hannah Hoag, Erik Ness, and Cassandra Willyard—gave critical early feedback and were among the book's first cheerleaders.

A number of people have laid eyes on the book, or portions of it, and helped make it whole. My daughter Grace Carpenter downloaded all the previously published articles from *The Open Notebook* and did much of the initial formatting. Aaron Brooks is the most thoughtful and fastidious copy editor I have ever worked with, and also the most fun. He has scoured every page of this book. (But since I fiddled with it after he did, any remaining errors are mine.) I'm also grateful to several extremely careful proofreaders: my aunt, Jane Considine; Jeanne Erdmann (again); my parents, Ann and Jake Delwiche; and my sister-in-law, Sophie Pierronnet. Lauren Gravitz helped me jump-start writing the book's introduction, after I had procrastinated doing so for 10 months, and then edited my first draft. Ann Finkbeiner, Julie Rehmeyer, and Alexandra Witze, all members of *TON*'s board, offered encouragement and wise counsel every step of the way, including providing crucial editorial help.

Who doesn't judge a book by its cover? My heartfelt thanks to Alison Mackey, the talented and science-obsessed art director who designed this book's cover, as well as its interior. I can't stop smiling about her work.

A number of people offered valuable technical help, for which I'm grateful. My brother, Luke Delwiche, helped me sort out a graphics issue that was tormenting me. Sarah Zielinski came to the rescue when my computer inexplicably could not print a PDF with a custom page size, and then she did it several times again when I kept making small changes. Michael Kranz, who didn't even know me, generously took time out of his day to troubleshoot another technical glitch. Gaius Augustus provided early help with graphic design, and I very much appreciate their time and creative ideas. Jim Webb, *TON*'s longtime and much-appreciated web developer, helped create the book's website. Eric Van Der Hope provided extensive assistance with the logistics of independent publishing. Kelly Tyrrell, the main voice of *TON* on social media, helps make sure that we keep in touch with our community and that when science writers speak, we hear them.

I'm grateful to the many friends, family members, and colleagues who have offered practical and moral support as this book has made its way from idea to completion, including: Christie Aschwanden, Stacey Baker, Sandra Beckwith, Bethany Brookshire, Jennifer Cutraro, Tinsley Davis, Francie Diep, Nadia Drake, Breanna Draxler, Rose Eveleth, Barbara Gastel, Cynthia Graber, Laura Helmuth, Jennifer Kahn, Andrea Ladd, Apoorva Mandavilli, Liz Neeley, Michelle Nijhuis, Claudia de Lima Oliveira, Rodrigo Pérez Ortega, Kendall Powell, Tina Hesman Saey, Jill Sakai, and Nicola Twilley. (Special thanks also to Nicola's beautiful, cover-inspiring water bottle.) I also thank the Binders Full of Science Writers community, whose members have cheered this book along, starting with a helpful discussion of various possible titles. Kyle Buchmann has acted as a combination of nonprofit management coach and psychotherapist for two years, and *TON* owes him a tremendous debt of gratitude.

I have benefited enormously from the expertise and generosity of Sarah Russo, who has offered masterful guidance about book publicity and marketing. If this book has the impact for science writers and aspiring science writers that I hope it will, it's thanks to the many hours Sarah spent graciously answering my questions—as well as the questions I didn't know to ask.

Finally, my deepest thanks and love to my husband, Joe Carpenter, who supports everything I do, and to my daughters, Grace and Alia, whom I adore and admire to a ridiculous degree.

∗

"How to Use Reporting Skills from Any Beat for Science Journalism" by Aneri Pattani. *TON*, April 24, 2018. © 2018 by Aneri Pattani.
"Trading the Pipette for the Pen: Transitioning from Science to Science Writing" by Julia Rosen. *TON*, June 16, 2015. © 2015 by Julia Rosen.

"Do You Need a Science Degree to Be a Science Reporter?" by Aneri Pattani. *TON*, August 21, 2018. © 2018 by Aneri Pattani.

"How to Break into English-Language Media as a Non-Native-English Speaker" by Humberto Basilio. *TON*, September 27, 2022. © 2022 by Humberto Basilio.

"Feeling Like a Fraud: The Impostor Phenomenon in Science Writing" by Sandeep Ravindran. *TON*, November 15, 2016. © 2016 Sandeep Ravindran.

"What Is Science Journalism Worth?" by Kendall Powell. Published in two parts, *TON*, January 20 and January 27, 2015. © 2015 by Kendall Powell.

"Nice Niche: How to Build and Keep Up with a Beat" by Knvul Sheikh. *TON*, June 4, 2019. © 2019 by Knvul Sheikh.

"A Conversation with Amy Maxmen about 'How the Fight against Ebola Tested a Culture's Traditions'" by Amanda Mascarelli. Published in "Storygram: Amy Maxmen's 'How the Fight against Ebola Tested a Culture's Traditions,'" *TON*, October 3, 2017. © 2017 by Amanda Mascarelli.

"Is This a Story? How to Evaluate Your Ideas Before You Pitch" by Mallory Pickett. *TON*, May 1, 2018. © 2018 by Mallory Pickett.

"Sharpening Ideas: From Topic to Story" by Dan Ferber. *TON*, July 11, 2012. © 2012 by Dan Ferber.

"Critically Evaluating Claims" by Megha Satyanarayana. *TON*, January 25, 2022. © 2022 by Megha Satyanarayana.

"Finding the Science in Any Story" by Kate Morgan. *TON*, November 27, 2018. © 2018 Kate Morgan.

"Pitching Errors: How Not to Pitch" by Laura Helmuth. *TON*, January 4, 2012. © 2012 by Laura Helmuth.

"Five Ways to Sink a Pitch" by Siri Carpenter. Not previously published. © 2020 by Siri Carpenter.

"What Makes a Good Pitch? Annotations from the TON Pitch Database" by Roxanne Khamsi. © 2020 by Roxanne Khamsi. Printed by permission of Roxanne Khamsi.

"A Conversation with Kathryn Schulz about 'The Really Big One'" by Michelle Nijhuis. Published as "Kathryn Schulz Paints a Chilling

Picture of 'The Really Big One,'" *TON*, September 15, 2015. © 2015 by Michelle Nijhuis.

"Is Anyone Out There? Sourcing News Stories" by Geoffrey Giller. *TON*, April 7, 2015. © 2015 by Geoffrey Giller.

"Interviewing for Career-Spanning Profiles" by Alla Katsnelson. *TON*, March 27, 2018. © 2018 by Alla Katsnelson.

"How to Conduct Difficult Interviews" by Mallory Pickett. *TON*, December 11, 2018. © 2018 by Mallory Pickett.

"Including Diverse Voices in Science Stories" by Christina Selby. *TON*, August 23, 2016. © 2016 by Christina Selby.

"How to Find Patient Stories on Social Media" by Katherine J. Wu. *TON*, August 11, 2020. © 2020 by Katherine J. Wu.

"Pulling It All Together: Organizing Reporting Notes" by Sarah Zhang. Not previously published. © 2020 by Sarah Zhang. Printed by permission of Sarah Zhang.

"Gut Check: Working with a Sensitivity Reader" by Jane C. Hu. *TON*, January 21, 2020. © 2020 by Jane C. Hu.

"When Science Reporting Takes an Emotional Toll" by Wudan Yan. *TON*, October 10, 2017. © 2017 by Wudan Yan.

"A Conversation with Annie Waldman on 'How Hospitals Are Failing Black Mothers'" by Tasneem Raja. Published in "Storygram: Annie Waldman's 'How Hospitals Are Failing Black Mothers,'" *TON*, March 19, 2019. © 2019 by Tasneem Raja.

"Good Beginnings: How to Write a Lede Your Editor and Your Readers Will Love" by Robin Meadows. *TON*, July 14, 2015. © 2015 by Robin Meadows.

"Nailing the Nut Graf" by Tina Casagrand Foss. *TON*, April 29, 2014. © 2014 by Tina Casagrand Foss.

"How to Find and Use Quotes in Science Stories" by Abdullahi Tsanni. *TON*, December 21, 2021. © 2021 by Abdullahi Tsanni.

"Like Being There: How Science Writers Use Sensory Detail" by Jyoti Madhusoodanan. *TON*, October 17, 2012. © 2012 by Jyoti Madhusoodanan.

"Eradicating Ableist Language Yields More Accurate and More Humane Journalism" by Marion Renault. *TON*, June 27, 2023. © 2023 by Marion Renault.

"Good Endings: How to Write a Kicker Your Editor and Your Readers Will Love" by Robin Meadows. *TON*, November 24, 2015. © 2015 by Robin Meadows.

"The First Critic Is You: Editing Your Own Work" by Tiên Nguyễn. *TON*, June 24, 2014. © 2014 by Tiên Nguyễn.

"A Conversation with Linda Nordling on 'How Decolonization Could Reshape South African Science'" by Jeanne Erdmann. Published as "Linda Nordling Probes the Transformation of South African Science," *TON*, July 10, 2018. © 2018 by Jeanne Erdmann.

"How to Read a Scientific Paper" by Alexandra Witze. *TON*, November 6, 2018. © 2018 by Alexandra Witze.

"What Are the Odds? Reporting on Risk" by Jane C. Hu. *TON*, November 1, 2016. © 2016 by Jane C. Hu.

"Spotting Shady Statistics" by Rachel Zamzow. *TON*, December 5, 2017. © 2017 by Rachel Zamzow.

"Problems with Preprints: Covering Rough Draft Manuscripts Responsibly" by Roxanne Khamsi. *TON*, June 1, 2020. © 2020 by Roxanne Khamsi.

"Getting the Most out of Scientific Conferences" by Rodrigo Pérez Ortega. *TON*, June 13, 2017. © 2017 by Rodrigo Pérez Ortega.

"Interrogating Data: A Science Writer's Guide to Data Journalism" by Betsy Ladyzhets. *TON*, July 28, 2020. © 2020 by Betsy Ladyzhets.

"Explaining Complexity" by Carl Zimmer. *TON*, July 7, 2015. © 2015 by Carl Zimmer.

"How to Do a Close Read" by Siri Carpenter. *TON*, March 13, 2018. © 2018 by Siri Carpenter.

"A Conversation with Maggie Koerth on 'The Complicated Legacy of a Panda Who Was Really Good at Sex'" by Ed Yong. Published as "Maggie Koerth Explores the Legacy of a Very Prolific Panda," *TON*, January 30, 2018. © 2018 by Ed Yong.

Contributors

HUMBERTO BASILIO is a Mexican freelance science writer based in New York City. He has written for *Eos*, *SciDev.Net*, *World Wildlife*, *Archeology*, and other publications. He is a member of the Mexican Network of Science Journalists and the Oxford Climate Journalism Network.

SIRI CARPENTER is cofounder, executive director, and editor-in-chief of *The Open Notebook*. She's an award-winning science journalist and editor whose writing and editorial work has appeared in the *New York Times*, *Science News*, *Science*, *Discover*, *bioGraphic*, *Scientific American*, and many other publications. She's a past president of the National Association of Science Writers.

TINA CASAGRAND FOSS is founder, editor, and executive director of *The New Territory*, a longform magazine about the Lower Midwest.

JEANNE ERDMANN is cofounder and editor-at-large of *The Open Notebook*, and vice chair of the board. She's an award-winning health and science journalist whose work has appeared in *Prevention*, *Discover*, *Real Simple*, *Family Circle*, *Women's Health*, *Slate*, the *Washington Post*, *Nature Medicine*, and *Nature*, among other publications.

DAN FERBER is a journalist whose work has appeared in *Science*, *Popular Science*, *Wired*, and many other outlets. A former long-time contributing correspondent for *Science*, he has worked as senior editor at *Discover* and at *Mechanical Engineering* magazine and is coauthor, with the late Harvard University public health expert Paul Epstein, MD, of *Changing Planet, Changing Health: How the Climate Crisis Threatens Our Health and What We Can Do about It*.

GEOFFREY GILLER worked as a freelance science writer for a decade, where his work appeared in the *New York Times, Discover, Scientific American, Audubon, bioGraphic,* and elsewhere. He is now a staff writer at the Yale University Office of Development.

LAURA HELMUTH is the editor-in-chief of *Scientific American,* and has been an editor for the *Washington Post, National Geographic, Slate, Smithsonian,* and *Science.* She has served on the advisory boards of several national publications and science organizations and is a past president of the National Association of Science Writers.

JANE C. HU is an independent journalist whose work has appeared in *Wired, National Geographic,* the *Atlantic, Outside, Slate,* and elsewhere. She is also a contributing editor at *High Country News,* and chair of *The Open Notebook*'s board of directors.

ALLA KATSNELSON is a freelance science writer and editor who specializes in biology, health and medicine, technology, and science policy. Her work has appeared in *Chemical & Engineering News, Knowable, Nature, BBC Focus,* the *New York Times,* and other outlets.

ROXANNE KHAMSI is a journalist whose writing has appeared in publications such as the *Economist, Scientific American, Slate,* and the *New York Times Magazine.* She served as chief news editor at *Nature Medicine* for more than a decade and taught science communication at the Alan Alda Center for Communicating Science at Stony Brook University and health journalism at the Craig Newmark Graduate School of Journalism at CUNY.

BETSY LADYZHETS is a science, health, and data journalist and writer focused on COVID-19 and the future of public health. She's coeditor of *The Sick Times,* a new nonprofit publication chronicling the Long COVID crisis, which she cofounded after concluding three years at the *COVID-19 Data Dispatch,* a newsletter and blog about tracking the pandemic. She also works part-time at the Council for the Advancement of

Science Writing and freelances for popular science outlets such as *Science News*, the *Atlantic*, and *STAT*.

JYOTI MADHUSOODANAN is a freelance science journalist. Her work has appeared in the *New York Times*, *Science*, *Nature*, *Scientific American*, and other outlets. She is a contributing writer at *Undark* magazine.

AMANDA MASCARELLI is the senior health and medicine editor at The Conversation US. Prior to that role, she was a 2020–2021 Ted Scripps Fellow in Environmental Journalism at the University of Colorado, Boulder, and was formerly the founding managing editor of *Sapiens*. Her work has been published in *National Geographic*, *Nature*, *Science*, the *New York Times*, the *Washington Post*, and elsewhere.

ROBIN MEADOWS is an independent journalist and a water reporter for *Maven's Notebook*, a California water news site. Her work has also appeared in *bioGraphic*, *Chemical & Engineering News*, *High Country News*, *Scientific American*, and elsewhere.

KATE MORGAN is a freelance journalist whose science and science-adjacent writing has appeared in the *New York Times*, the *Washington Post*, the *Wall Street Journal*, *Sierra*, *National Geographic*, *Harper's Bazaar*, *Slate*, *Saveur*, and many other publications.

TIÊN NGUYỄN is an independent journalist, documentary filmmaker, and reformed PhD chemist. Her stories span science, history, and culture and have appeared in *Chemical & Engineering News*, *Nature*, *Science*, *Scientific American*, and *Vice News*. She has also produced short documentaries for PBS Digital Studios, KCET, ITVS, and *Knowable*.

MICHELLE NIJHUIS is an award-winning journalist and the author of *Beloved Beasts: Fighting for Life in an Age of Extinction*, a history of the modern conservation movement. A longtime contributing editor of *High Country News* and a regular contributor to the *New York Review of Books*, she has also reported on science and the environment for *Na-*

tional Geographic, the *Atlantic*, and the *New York Times Magazine*. She is coeditor of *The Science Writers' Handbook* and author of *The Science Writers' Essay Handbook*.

ANERI PATTANI is a senior correspondent for *KFF Health News*, a national nonprofit covering US health care and health policy. Pattani reports on a range of public health topics, with a focus on mental health, suicide, and substance use. Her work spans text and audio stories, and she has been heard on NPR and *Science Friday*. She was a 2019 recipient of the Rosalynn Carter Fellowship for Mental Health Journalism. She is pursuing her master's degree in public health as a Bloomberg fellow at Johns Hopkins University.

RODRIGO PÉREZ ORTEGA is an award-winning science journalist and a staff writer for *Science*. His work has also appeared in *Nature*, the *New York Times*, *El País*, and other publications. He is a founding member of the Mexican Network of Science Journalists and is the editorial director for *The Open Notebook*'s Spanish translations program.

MALLORY PICKETT is a science writer and editor who has written about science, technology, and the environment for the *New York Times Magazine*, *Wired*, *FiveThirtyEight*, the *Guardian*, and other publications. She's now the head of editorial at Watershed.

KENDALL POWELL is a freelance science writer and editor whose work has appeared in the *Washington Post*, *Nature*, *Knowable*, *Mosaic*, and other publications. She's a contributor to *The Science Writers' Handbook: Everything You Need to Know to Pitch, Publish and Prosper in the Digital Age*. She's currently a senior editor at *Nature Careers*.

TASNEEM RAJA is editor-in-chief of of the *Oaklandside*, a local newsroom serving Oakland, California. She's also cofounder of the *Cityside Journalism Initiative*, the nonprofit parent of the *Oaklandside* and *Berkeleyside*. Previously, she was executive editor of the *Tyler Loop*, which explores policy, history, and demographics in Tyler, Texas. She is an award-winning journalist who has reported for NPR, the *New Yorker*, the *Atlantic*, *Mother Jones*, and other national outlets.

SANDEEP RAVINDRAN is a freelance science writer who has written about life sciences and technology for publications such as *Smithsonian*, *National Geographic News*, *Nature*, the *Scientist*, *Wired*, *Popular Science*, and *Science News for Students*. He is a program manager at *The Open Notebook* and is president of the National Association of Science Writers.

MARION RENAULT is a freelance science and health writer whose work has appeared in the *Atlantic*, the *New Republic*, *Slate*, *STAT*, the *New Yorker*, *Wired*, the *New York Times*, and elsewhere.

JULIA ROSEN is a freelance journalist whose work has appeared in the *Atlantic*, the *New York Times*, *Science*, *Nature*, and *High Country News*, among other outlets. She is a former science reporter at the *Los Angeles Times*.

MEGHA SATYANARAYANA is the chief opinion editor at *Scientific American*, where she works with writers of all scientific stripes to publish their expert thoughts. She has been a reporter and editor at *Chemical & Engineering News* and *STAT* and has worked at several daily newspapers since graduating from the UC Santa Cruz Science Communication Program. She has a PhD in molecular biology.

CHRISTINA SELBY is a freelance writer and conservation photographer whose work has appeared in *bioGraphic*, *Audubon*, *National Geographic* online, *Sierra Magazine*, *Scientific American*, *New Mexico Magazine*, *High Country News*, and other publications. She is also the author of *New Mexico Wildflower Hiking Guide* and *Family Outdoor Adventures New Mexico*.

KNVUL SHEIKH is a *New York Times* health and wellness reporter. Her work has appeared in the *Atlantic*, *National Geographic*, *Scientific American*, *Scholastic News*, and elsewhere.

ABDULLAHI TSANNI is a science writer specializing in narrative features and public-interest science journalism. He covers biotechnology and biomedical research, global health, AI, and the scientific enterprise. His work has appeared in *Nature, STAT, MIT Technology Review,* the *British Medical Journal,* and other publications.

ALEXANDRA WITZE is a freelance journalist whose work has appeared in *Nature, Science News, Knowable,* and many other outlets. With her husband Jeff Kanipe, she is coauthor of *Island on Fire,* a book about the 18th-century eruption of the Icelandic volcano Laki. She is on the board of *The Open Notebook.*

KATHERINE J. WU is a staff writer at the *Atlantic,* a senior editor at *The Open Notebook,* and a senior producer for the *Story Collider.* She previously served as a science reporter for the *New York Times.* She won a Schmidt Award for Excellence in Science Communication in 2022, a Science in Society journalism award in 2021, and the Evert Clark/Seth Payne Award for Young Science Journalists in 2020. She has a PhD in microbiology from Harvard University.

WUDAN YAN is an award-winning narrative journalist, fact-checker, and entrepreneur. Her writing has appeared in the *Atlantic, California Sunday Magazine, High Country News, MIT Technology Review,* the *New York Times, Popular Mechanics,* and elsewhere. She is also the founder and executive producer of the Writers' Co-op, a business podcast and learning academy for freelance creatives.

ED YONG is a science writer and the author of two *New York Times* bestsellers: *An Immense World* and *I Contain Multitudes.* He won the Pulitzer Prize in Explanatory Journalism for his coverage of the COVID-19 pandemic.

RACHEL ZAMZOW is an award-winning science journalist and editor based in central Texas. She's the managing editor of *The Open Notebook.* And she reports on neuroscience and science ethics for *Science, Science News, Spectrum,* and other outlets.

SARAH ZHANG is a staff writer at the *Atlantic*, where she covers science. She was previously a staff writer at *Wired*, and her work has also appeared in the *New York Times*, *Nature*, and *Discover*.

CARL ZIMMER is a journalist and author of 14 books about science, including *She Has Her Mother's Laugh: The Powers, Perversions, and Potential of Heredity*, which won the 2019 National Academies Communication Award, among other honors. He is also the author of the Matter column in the *New York Times* and is professor adjunct in the Department of Molecular Biophysics and Biochemistry at Yale University, where he teaches science writing.

Index

**Chicago Guides
to Writing, Editing,
and Publishing**

A complete list of series titles is available on the University of Chicago Press website.

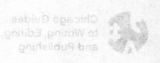

Printed and bound by CPI Group (UK) Ltd, Croydon, CR0 4YY

16/04/2025

14658566-0003